OCCASIONAL PAPER

The Perfect Storm

The Goldwater-Nichols Act and Its Effect on Navy Acquisition

Charles Nemfakos • Irv Blickstein • Aine Seitz McCarthy • Jerry M. Sollinger

Prepared for the United States Navy

RAND NATIONAL DEFENSE RESEARCH INSTITUTE

The research described in this report was prepared for the United States Navy. The research was conducted in the RAND National Defense Research Institute, a federally funded research and development center sponsored by the Office of the Secretary of Defense, the Joint Staff, the Unified Combatant Commands, the Navy, the Marine Corps, the defense agencies, and the defense Intelligence Community under Contract W74V8H-06-C-0002.

Library of Congress Cataloging-in-Publication Data

The perfect storm : the Goldwater-Nichols Act and its effect on Navy acquisition / Charles Nemfakos ... [et al.].
    p. cm.
  Includes bibliographical references.
  ISBN 978-0-8330-5018-2 (pbk. : alk. paper).
  1. United States. Goldwater-Nichols Department of Defense Reorganization Act of 1986. 2. United States.
Navy—Procurement. 3. United States. Dept. of Defense—Reorganization. 4. United States—Armed Forces—
Reorganization—History. 5. Military law—United States—History. I. Nemfakos, Charles.

    KF7252.P47 2010
    343.73'0196212—dc22

                                                                                        2010028209

The RAND Corporation is a nonprofit institution that helps improve policy and decisionmaking through research and analysis. RAND's publications do not necessarily reflect the opinions of its research clients and sponsors.

RAND® is a registered trademark.

Published 2010 by the RAND Corporation
1776 Main Street, P.O. Box 2138, Santa Monica, CA 90407-2138
1200 South Hayes Street, Arlington, VA 22202-5050
4570 Fifth Avenue, Suite 600, Pittsburgh, PA 15213-2665
RAND URL: http://www.rand.org/
To order RAND documents or to obtain additional information, contact
Distribution Services: Telephone: (310) 451-7002;
Fax: (310) 451-6915; Email: order@rand.org

Preface

In 1986, beginning with the Goldwater-Nichols Department of Defense Reorganization Act, the U.S. military establishment underwent the most sweeping set of defense reforms to be enacted in almost 40 years. Related reforms followed shortly thereafter and included those contained in the National Defense Authorization Act of 1987, which reflected many of the recommendations of the Packard Commission.[1] In the more than two decades since that time of change, the military establishment has taken numerous steps to implement the legislation's reforms and other reforms contained in commission recommendations and further legislation. Although reform was necessary, some within the military services have grown increasingly concerned about some of the effects, perceiving a growing divide between a military-run requirements process and a civilian-run acquisition process—a divide they regard as inimical to the efficient and effective support of military forces.

This paper focuses on the implementation of the Goldwater-Nichols Act in the Department of the Navy and on related acquisition reforms, but it also assesses the influence of several other factors that, in large part, made passage of the Goldwater-Nichols legislation possible, colored its implementation, and complicated the adoption of common-sense changes during the implementation process.

To thoroughly examine both specific issues and general conditions, the RAND Corporation undertook both objective research and an interview process that engaged former officials who served during the initial enactment of the Goldwater-Nichols and acquisition reform legislation. The two principal authors of this paper were senior officials within the Department of the Navy (DoN) during the implementation of the sweeping changes resulting from the various acts, they participated personally in the promulgation of many applicable internal DoN regulations, and they were familiar with the principal actors on the scene during the mid- to late 1980s and early 1990s. Their knowledge and insight has enabled the interweaving of the research and interview processes. We attempt to highlight where objective research and interviews and personal insights are the predominant basis for comments and observations. When not otherwise indicated, the paper is informed by all three.

Because of the nature of the convergent issues and the approach taken by the authors, the paper should be of interest to the Department of Defense–wide acquisition community, requirements-generating offices in the three military departments, students of large public policy shifts in defense in the 1980s and 1990s, and members of the U.S. Congress.

---

[1]   David Packard, *President's Blue Ribbon Commission on Defense Management, A Quest for Excellence: Final Report to the President*, Washington, D.C., June 30, 1986.

This research was sponsored by the Naval Sea Systems Command at the behest of the Chief of Naval Operations and the Office of the Secretary of Defense for Acquisition, Technology and Logistics and conducted within the Acquisition and Technology Policy Center of the RAND National Defense Research Institute, a federally funded research and development center sponsored the Office of the Secretary of Defense, the Joint Staff, the Unified Combatant Commands, the Navy, the Marine Corps, the defense agencies, and the defense Intelligence Community.

For more information on the RAND Acquisition and Technology Policy Center, see http://www.rand.org/nsrd/about/atp.html or contact the Director (contact information is provided on the web page).

Contents

Figures

Tables

Summary

The Goldwater-Nichols Department of Defense Reorganization Act passed in 1986 was one of the most sweeping pieces of legislation to affect the Department of Defense and the military services in decades. Its passage resulted from dissatisfaction on the part of Congress and other influential policymakers with what they perceived as the U.S. military's stubborn refusal to deal with long-festering problems. These problems included an inability on the part of the military services to mount effective joint operations and an inefficient, unwieldy, and at times corrupt system for acquiring weapon systems. These perceptions had some basis in reality. The historical landscape was littered with examples of mishandled military operations, including the Vietnam War and the failed attempts to rescue both the crew of the SS *Mayaguez* and the Americans taken hostage in Iran. The acquisition process fared no better in terms of success, as proven by the Ill Wind investigation, huge cost overruns, and such flawed systems as the A-12 Avenger.

But Goldwater-Nichols was only one manifestation of widespread discontent with the Department of Defense's operational and acquisition capabilities. Between 1986 and 1990, a remarkable number of events changed how the department was organized, conducted military operations, and did business. The climate surrounding the enactment of Goldwater-Nichols was indeed a "perfect storm," a confluence of disparate currents, some flowing from long-standing problems and others from more-recent events. These currents not only facilitated the passage of Goldwater-Nichols but also shaped its implementation in the military departments.

This paper focuses on the implementation of Goldwater-Nichols in DoN. It argues that the implementation of the act in DoN had three undesirable consequences:

- It erected an impenetrable wall between a military-controlled requirements process and a civilian-driven acquisition process to the overall detriment of acquisition in DoN.
- Its personnel policies deprived the DoN of a blended acquisition workforce composed of line officers with extensive operational experience who provided valuable perspective that those who spent most of their careers in acquisition assignments lacked.
- It created a generation of line officers who had little or no understanding of or appreciation for the acquisition process.

These consequences were unintended by those who crafted the legislation but were exacerbated by DoN's overly restrictive interpretation of the legislation.

To rectify the situation, we recommend that DoN

- change its directives to eliminate the wall between the requirements and acquisition processes and spell out a continuing role for the Chief of Naval Operations and the Commandant of the Marine Corps that is more in line with the practices of the other military services
- create an acquisition oversight body co-chaired by the Assistant Secretary of the Navy for Research, Development and Acquisition; the Vice Chief of Naval Operations; and, in matters of priority interest to the Marine Corps, the Assistant Commandant of the Marine Corps
- create desirable career opportunities for line officers in the material establishment.

In the final analysis, institutional balance is a central element of concern. Violent storms disturb the evolved balance of nature, with that equilibrium being restored over time. The authors observe that the "perfect storm" addressed in this paper distorted the balance of actors and forces that was key to institutional governance. A quarter of a century later, that balance has not been regained; if anything, distortions continue. The recommendations are a step in restoring that institutional balance.

Acknowledgments

Many individuals and interviewees contributed to our study, and we would like to thank them. First and foremost, we would like to thank Brian Persons of Naval Sea Systems Command 05B for both his sponsorship of this study and for providing input and guidance along the way. We would also like to thank the various interviewees who took time out of their busy schedules to meet with us. Without their insight and experiences, documentation of the important events described in this paper would not have been possible. We apologize if we did not fully capture their insights. To others whose insights could have contributed to our efforts but whom we were unable to reach or whom we simply did not think of, we apologize as well.

This report has benefitted from the reviews of Jeff Drezner and Susan Marquis, both of RAND. Jennifer Miller's advice was invaluable during revisions.

Any inaccuracies or misunderstanding of the history of this period are the responsibility of the authors alone.

Abbreviations

AAE	Army Acquisition Executive
ACAT	acquisition category
AFAE	Air Force Acquisition Executive
AFHQ	Air Force Headquarters
AFI	Air Force Instruction
AFMC	Air Force Materiel Command
AFR	Air Force Regulation
AFSARC	Air Force Systems Acquisition Review Council
AFSC	Air Force Systems Command
AMC	Army Materiel Command
AR	Army Regulation
ASA	Assistant Secretary of the Army
ASA (ALT)	Assistant Secretary of the Army for Acquisition, Logistics and Technology
ASA (RDA)	Assistant Secretary of the Army for Research, Development and Acquisition
ASAF (A)	Assistant Secretary of the Air Force for Acquisition
ASARC	Army Systems Acquisition Review Council
ASN	Assistant Secretary of the Navy
ASN (FM)	Assistance Secretary of the Navy for Financial Management
ASN (RD&A)	Assistant Secretary of the Navy for Research, Development and Acquisition
ASN (RE&S)	Assistant Secretary of the Navy for Research, Engineering and Systems
ASN (S&L)	Assistant Secretary of the Navy for Shipbuilding and Logistics

CAE	Component Acquisition Executive
CBTDEV	combat developer
CEB	CNO Executive Board
CINC	Commander-in-Chief
CMC	Commandant of the Marine Corps
CNO	Chief of Naval Operations
CSA	Chief of Staff of the Army
CSAF	Chief of Staff of the Air Force
DAE	defense acquisition executive
DARCOM	Development and Readiness Command
DAU	Defense Acquisition University
DCNO	Deputy Chief of Naval Operations
DCS	Deputy Chief of Staff
DCSOPS	Deputy Chief of Staff for Operations and Plans
DCSRDA	Deputy Chief of Staff for Research, Development and Acquisition
DMR	Defense Management Review
DoD	Department of Defense
DoDD	Department of Defense Directive
DoDI	Department of Defense Instruction
DoN	Department of the Navy
DOTLMPF	doctrine, organization, training, materiel, leadership and education, personnel, and facilities
GAO	General Accounting Office
GRH	Gramm-Rudman-Hollings
HQDA	Headquarters, Department of the Army
MATDEV	material developer
NAE	Navy Acquisition Executive
NDAA	National Defense Authorization Act for Fiscal Year 1987
NMC	Naval Materiel Command
OPNAV	Office of the Chief of Naval Operations
OSD	Office of the Secretary of Defense

PEO	program executive officer
PM	program manager
POM	Program Objectives Memorandum
PPBE	Planning, Programming, Budgeting, and Execution
PPBES	Planning, Programming, Budgeting, and Execution System
SAE	service acquisition executive
SAF/AL	Assistant Secretary of the Air Force for Research, Development and Logistics
SAF/AQ	Assistant Secretary of the Air Force for Acquisition
SAF/US	Under Secretary of the Air Force
SecAF	Secretary of the Air Force
SECNAV	Secretary of the Navy
SECNAVINST	Secretary of the Navy Instruction
SECNAVNOTE	Secretary of the Navy Notice
SPE	senior procurement executive
SYSCOM	Systems Command
TRADOC	Training and Doctrine Command
TNGDEV	training developer
USD (A)	Undersecretary of Defense for Acquisition
USD (AT&L)	Undersecretary of Defense for Acquisition, Technology and Logistics
VCNO	Vice Chief of Naval Operations

CHAPTER ONE

Introduction

The debate over the appropriate roles of the Chief of Naval Operations (CNO) and of the Secretary of the Navy (SECNAV) in the material management process stretches back to the Civil War era.[1] The essence of the debate is the role of uniformed leadership (i.e., CNO) compared with that of civilian leadership (i.e., SECNAV) in determining what warfighting capabilities are required, what systems will be procured to provide these capabilities, how these systems will be supported when introduced into the fleet, and how these systems will be funded. In 1986, the Goldwater-Nichols Department of Defense Reorganization Act (P.L. 99-433) weighed in on these roles as a key element in its overall reform of defense organization and processes, giving responsibility for defense acquisitions to civilian secretaries while strengthening joint uniformed oversight over the requirements process.

Since the enactment of this momentous legislation, the military services have taken numerous steps to implement its provisions and to respond to related acquisition reforms. However, some senior Navy officials have grown increasingly concerned about the unintended consequences of these reforms, perceiving a growing divide between a military-run requirements process and a civilian-run acquisition process.

Objectives and Approaches

RAND examined (1) the operational, budgetary, and policy issues that drove the passage of the Goldwater-Nichols Act and related acquisition reforms and (2) the Department of the Navy's (DoN's) implementation of these reforms, particularly with regard to their influence on military and civilian roles in the DoN's acquisition process. This paper describes the context in which acquisition reform occurred and the effects of that reform on acquisition processes, focusing largely on DoN. Drawing on a series of interviews with numerous officials who were present when the legislation was implemented, we conclude that the reform's effect was to focus the CNO's attention on requirements issues and to divorce the position from the acquisition process in a way that has been detrimental to the effective and efficient acquisition of material for DoN. It further argues that this separation went beyond what the legislation required and that there needs to be closer integration of CNO's interests with those of the Assistant Secretary of the Navy (ASN) for Research, Development and Acquisition (ASN (RD&A)) and

---

[1] Edwin Hooper and Thomas Hone have written richly detailed, historical examinations of this debate. See Edwin B. Hooper, *The Navy Department: Evolution and Fragmentation*, Washington, D.C.: Naval Historical Foundation, 1978; Thomas C. Hone, *Power and Change: The Administrative History of the Office of the Chief of Naval Operations, 1946–1986*, Washington, D.C.: Naval Historical Center, 1989.

of the Navy acquisition community to increase material capabilities and readiness at reduced cost.

This paper deals with more than the Goldwater-Nichols legislation and considers several other influences, such as the troubled history of the armed forces in coordinating joint operations and the effect of such significant commissions as the Packard Commission. These other influences coalesced in the mid-1980s and created an environment—a perfect storm[2]—that both made the passage of Goldwater-Nichols possible and colored its implementation. In essence, the Goldwater-Nichols legislation stands as a proxy for these other influences.

To understand the policy issues behind the Goldwater-Nichols Act and related acquisition reforms, we reviewed literature on the political and economic environment leading up to these initiatives and examined analyses of defense acquisition problems.[3] To understand how DoN implemented acquisition reforms and the effect of this implementation, we reviewed DoN implementation guidance[4] and Department of Defense (DoD) guidance,[5] and we interviewed both former and current DoN civilian and military officials and civilian officials outside of DoN who were deeply involved in implementing Goldwater-Nichols and related reforms. A list of the positions formerly held by these individuals is provided in the appendix.

We also interviewed former Army and Air Force senior uniformed and civilian officials to compare both the implementation of Goldwater-Nichols in those departments and services and the influences of other reforms with implementation and influences in DoN. We have attempted to capture and present a synthesis of their views.

We recognize the inherent limitations in this approach: For example, we interviewed only a very small subset of the many people involved over the years, and those we interviewed provided their recollections of events that occurred more than 20 years ago. That said, those we interviewed were key players during the implementation, and they reported firsthand experiences. Also, because they were interviewed separately, we were able to crosscheck each account with the others. Furthermore, much of our discussion with the interviewees concerned the effects of implementation, and the interviewees were uniquely qualified to analyze both the legislation's effect on processes and the implications of the divide between the requirements and acquisition processes.

How the Paper Is Organized

Chapter Two describes the context surrounding the legislation, including the historical context of military operations and the events pertaining to acquisition practice in DoD. Further,

---

[2]   The phrase *perfect storm* is used to describe an event where a rare combination of circumstances exacerbates a situation drastically. It was also the title of a 1997 book and a 2000 movie adapted from the book.

[3]   We reviewed Dick Cheney, *Defense Management Report to the President*, Washington, D.C.: Department of Defense, 1989; the Packard Commission Report (David Packard, *President's Blue Ribbon Commission on Defense Management, A Quest for Excellence: Final Report to the President*, Washington, D.C., June 30, 1986); the Joint Defense Capabilities Study Team, *Joint Defense Capability Study: Improving DoD Strategic Planning, Resourcing and Execution to Satisfy Joint Capabilities*, Washington, D.C.: Department of Defense, 2004; Assessment Panel of the Defense Acquisition Performance Assessment Project, *Defense Acquisition Performance Assessment: Executive Summary*, December 2005; and assessments conducted by the Center for Strategic and International Studies and the Government Accountability Office.

[4]   Such as SECNAVINST 5400.15C.

[5]   Such as DoDI 5000.02, in its multiple iterations.

it describes the passage of the Goldwater-Nichols Act, the main players in its enactment, and its key provisions, and it discusses the National Defense Authorization Act for Fiscal Year 1987 (P.L. 99-961, abbreviated as NDAA). Chapter Three examines the passage of the act and how it was implemented in the DoN. Chapter Four describes the military services' military acquisition process both before and after the Goldwater-Nichols legislation, noting the key changes that occurred in service acquisition practices as a result of that legislation. Chapter Five describes how Goldwater-Nichols and the NDAA affected the acquisition process in the DoN. Chapter Six presents our conclusions and some suggested courses of action for the DoN to improve its ability to acquire equipment, and it identifies areas that warrant further study.

The Context of Goldwater-Nichols

The passage of the Goldwater-Nichols Act in 1986 resulted from operational, organizational, and fiscal pressures that had been building for a number of years and, indeed, continued after the act was passed. These pre- and postenactment events are important because they provide the context in which legislation was passed and implemented in DoD and the military services. This chapter briefly describes these pressures and events and their significance in the crafting, passage, and implementation of the legislation.

Timeline

Figure 2.1 portrays the timeline of events that occurred before, during, and after the passage of Goldwater-Nichols. The timeline underscores several points. First, the forces that eventually called Goldwater-Nichols into being began to arise in the decades before the act was passed. Second, these forces manifested themselves in quite different venues: in the operational performance of U.S. military forces, in the performance of the system that governed the acquisition of military weapons and weapon systems, and in the behavior and practices of those who operated in that system. Third, a remarkable number of important events that occurred between 1985 and 1990 built an almost unstoppable momentum that ensured that long-standing issues would finally be dealt with in a systematic way. The effect of the whole far exceeded the power of the individual parts. The following sections briefly describe the events that contributed to the eventual perfect storm.

Operational Shortcomings

A series of either failed or less-than-satisfactory military operations sparked among policymakers widespread discontent with the performance of the U.S. military. Further fueling that discontent was the fact that the problems appeared to be systemic and not simply the failure of any particular set of individuals. After-action reports called attention to, among other things, the failure of the military services to work together as a harmonious whole. The problem persisted, and policymakers eventually lost faith in the will or the ability of the military services to overcome service parochialism in the interest of developing joint capabilities.

Figure 2.1
Events Contributing to the Context of Goldwater-Nichols

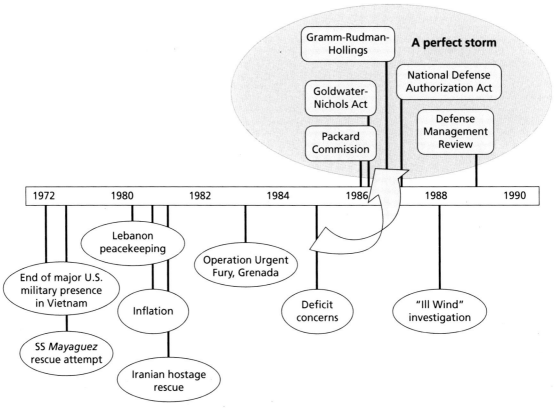

RAND OP308-2.1

Vietnam

Both the conduct of the Vietnam War and, in particular, its outcome caused many in the United States, including influential legislators, to voice dissatisfaction with the effectiveness of DoD. Among these many critics was defense analyst Jeffrey Record, who, expressing what ultimately became widespread opinion about the major cause of the United States' defeat, laid the blame squarely at the feet of military and government leaders, citing their failure to establish clear objectives, their unwarranted faith in ground combat operations to counter an insurgency, and their misreading of the staying power of the North Vietnamese.[1] Other postmortems noted the lack of leadership by the Joint Chiefs of Staff and the corrosive effect of service parochialism.[2] The sentiment of the American public during the post-Vietnam era was one of general discontent and disillusionment with all things military. These feelings were reflected in the views held by the members of Congress. Later events, such as those described in the next sections, only intensified these feelings and convinced key members of Congress that the military was incapable of taking the necessary steps to correct the problems on its own.

---

[1]   Jeffrey Record, *The Wrong War: Why We Lost in Vietnam*, Annapolis, Md.: Naval Institute Press, 1998.

[2]   See, for example, H. R. McMaster, *Dereliction of Duty: Lyndon Johnson, Robert McNamara, the Joint Chiefs of Staff, and the Lies That Led to Vietnam*, New York: Harper-Collins, 1997.

The SS *Mayaguez* Rescue Attempt

In May 1975, two weeks after the fall of Saigon to the forces of North Vietnam, the Khmer Rouge government of Cambodia seized a U.S. container ship, the SS *Mayaguez*, in international waters (that Cambodia claimed as its own) and removed the crew for questioning. Having just suffered the ignominy of defeat in Vietnam, the United States felt the need to make a decisive statement and opted to stage a military rescue of the crew. The ensuing operation involved Navy, Marine Corps, and Air Force elements along with various intelligence agencies. The central operational action was an assault on an island off the coast of Cambodia where the United States thought the crew was being held captive. In reality, however, the crew had been released before the operation began, and the island had been heavily fortified for reasons that had nothing to do with the SS *Mayaguez*. Intelligence agencies failed to detect the fortifications. The rescue operation was a debacle that resulted in approximately 20 killed and 40 wounded U.S. personnel. Exacerbating this unsought outcome was the fact that several of the service members' bodies were not retrieved and that three marines were left behind to be captured, tortured, and eventually killed. The U.S. military was criticized for faulty intelligence, coordination and communications problems, and the ad hoc and haphazard planning of the joint operation.[3]

Lebanon

Poor communication and unclear chains of command were similarly highlighted in analyses of the United States' involvement in a multinational peacekeeping force in Lebanon during the early 1980s.[4] One analysis determined that six command chains controlled the actions of the marine amphibious unit deployed in Lebanon and that this led to communication breakdowns and unclear authority.[5] Political and military leaders also frequently changed the rules of engagement during the two years when military forces were involved, leading to inconsistencies that may have created an atmosphere of hesitancy among marines that contributed to the death of 275 marines in a suicide bombing.

Grenada and Operation Urgent Fury

In 1983, when U.S. forces invaded the island of Grenada during Operation Urgent Fury (an effort to oust the Cuban-sponsored People's Revolutionary Government and to protect U.S. citizens on the southern Caribbean island), the same issues and operational failures arose: Poor communications, command and control, and planning diminished operational efficiency, despite the efforts of the troops on the ground.[6] For example, different force components—including rangers, paratroopers, sailors, and marines—lacked interoperable communication devices, and key players were excluded from planning, which led to problems with logistics

[3] See Ralph Wetterhahn, *The Last Battle: The Mayaguez Incident and the End of the Vietnam War*, New York: Carroll and Graf, 2001.

[4] Peter J. Ferraro, *Beirut, Lebanon: 24th MAU, May–Dec 1983*, Decatur, Ga.: Marine Corps University Command and Staff College, 1997.

[5] DoD Commission, *Report of the DoD Commission on Beirut International Airport Terrorist Act*, Washington, D.C.: Department of Defense, 1983.

[6] Stephen E. Anno and William E. Einspahr, "The Grenada Invasion," in Stephen E. Anno and William E. Einspahr, *Command and Control Lessons Learned: Iranian Rescue, Falklands Conflict, Grenada Invasion, Libya Raid*, Maxwell Air Force Base, Ala: Air University Press, 1988.

support.[7] The Office of the Chairman of the Joint Chiefs of Staff's lessons-learned report on the operation outlined the need for a more collaborative planning process among the services.[8] This report more clearly outlined planning for operations, identifying rapid responses as the most necessary and suggesting the pursuit of significant improvements. For example, it recommended the use of simple command structures with authority delegated to the lowest possible level.

The Iranian Hostage Rescue

The attempt to rescue American hostages from the U.S. embassy in Iran—Operation Eagle Claw—was a complex, two-day operation involving forces from the Army, the Navy, and the Air Force; operatives from the Central Intelligence Agency; and participants from other intelligence agencies. The plan called for U.S. forces to make a covert landing in C-130 aircraft at a temporary airstrip dubbed Desert One, join up with Navy helicopters, and redeploy to a hidden site from which the rescue operation would be launched the following night. Notably, the participating service units trained separately, meeting for the first time at Desert One. One participant described the meet-up: "four commanders at the scene without visible identification, incompatible radios, and no agreed-upon plan, not even a designated location for the commander."[9] Bad weather, helicopter maintenance problems, and a crash at the refueling site caused the command authority to abort the mission. Eight servicemen were killed in the crash.

The Holloway Report, which reported the results of a review of Operation Eagle Claw and was conducted for the Joint Chiefs of Staff and led by former CNO Admiral James Holloway III, directed much of its criticism for the failed hostage rescue mission toward the lack of joint training and coordination; the lack of integrated intelligence for use by the joint task force; overly complex, service-unique planning by each military service; and communication deficiencies.[10] The report also concluded that planning and coordination had been delayed because the Joint Chiefs of Staff had to start from scratch in creating a joint task force. The report recommended that each of these problems be addressed in future special operations and specifically recommended the formation of a Counterterrorist Joint Task Force and a Special Operations Advisory Panel.

Summary of Military Operational Deficiencies

This failed rescue attempt examined in the Holloway Report is a microcosm of all the problems encountered in what seemed to be an unending series of military disasters that shared similar shortcomings: muddled and multiple chains of command, poor interservice planning and coordination, ad hoc responses to each new crisis, the inability of one service to communicate with another, and interservice rivalries and parochialism that hampered the services' ability to work in concert. Furthermore, the military services did not seem to be able or even willing to resolve the problems.

---

[7] Anno and Einspahr, 1988.

[8] Joseph P. Doty, *Urgent Fury—A Look Back . . . A Look Forward*, Newport, R.I.: Naval War College, 1994.

[9] James R. Locher III, "Has It Worked?: The Goldwater-Nichols Reorganization Act," *Naval War College Review*, Vol. LIV, No. 4, 2001, pp. 95–114.

[10] James L. Holloway, *Special Operations Review of Iranian Hostage Rescue Mission, Joint Chiefs of Staff*, Washington, D.C., August 23, 1980.

These shortcomings were symptoms of structural problems that had been festering for decades. The general perception was that the existing national security structure promoted the interests of the services over those of the nation. Drawing on his experience as an Air Force commander in Europe, General David Jones offered an example of the nature of the problem:

> When I was the Air Commander in Europe, I had two bosses, the Chief of Staff of the Air Force and the Unified Commander—the Commander-in-Chief, U.S. European Command who is over all U.S. theater forces. The Chief of Staff of the Air Force assigned me all my people, gave all my rewards to my people, controlled all my money, gave me all my equipment. Obviously, he had nine times the influence over me than my Unified Commander had. So, he who controls the resources can have a tremendous impact.[11]

The failures described in this section, and the attendant loss of life of U.S. service personnel, led influential congressmen—notably, Senators Sam Nunn of Georgia and Barry Goldwater of Arizona—to use legislation as a way of forcing the military services to deal with the problems that seemed to stand in the way of effective joint operations. Although these problems were not directly caused by or connected to acquisition issues, some of the personnel actions directed by the legislation with an eye toward increasing the joint experience and expertise of the officer corps spilled over into the acquisition arena in ways that the legislation's sponsors likely did not intend.

Acquisition Shortcomings

Bungled military operations were not the only thing that fueled the fires of reform. Corrupt and inefficient acquisition processes, described in the following sections, also contributed.

Fraud

In a two-year investigation begun in 1986 and known as "Ill Wind," the Federal Bureau of Investigation investigated corruption in government and contractor interactions in acquisition programs. The investigation issued 250 subpoenas for evidence about the activities of more than 50 private consultants, a dozen defense-industry companies, and many DoD officials. Convictions resulting from the investigation included ASN (RD&A) and others charged with influence-peddling and leaking government information to defense firms. The investigation showed that improper contracting, fraud, and abuse were enabled by the restricted flow of accurate information up the chain of command, by the lack of financial and scheduling realism in programs, and by the perceived need to oversell programs to win a defense contract.[12]

---

[11] Committee on Armed Services, House of Representatives, "Background Material on Structure Reform of the Department of Defense," 99th Congress (2nd Session), Washington, D.C.: U.S. Government Printing Office, 1986, p. 5.

[12] An administrative inquiry directed by SECNAV noted similar contracting problems when its results were released in 1990. The memorandum forwarding the results of the inquiry to SECNAV concluded that the schedule and cost goals for the A-12 Avenger, a carrier-based stealth bomber designed to replace the A-6 Intruder, were overly optimistic and should not have been supported by government managers in the contract and program offices. It also concluded that those responsible for determining the costing of programs should make greater use of the Cost Schedule Control System, contract performance management, the Defense Acquisition Executive Summary, and earned value analysis. In addition, it noted that direct lines of authority needed to be established for joint government and contractor teams and that communication between them needed to be encouraged. The A-12 program was cancelled by Secretary of Defense Dick Cheney when its

After the "Ill Wind" investigation had run its course, a General Accounting Office (GAO) report on defense weapons system acquisition provided conclusions and specific action recommendations for DoD, addressing, among other things, financial realism, a freer flow of accurate information as a mechanism for limiting fraud and abuse, alignment of career success with better program outcomes, and prevention of the improper influencing of contract awards.[13] Some of the recommendations concerned, albeit in different terms, the intelligence and planning considerations relevant to the operational problems described earlier in this chapter. The cumulative effect of this and other reports, the procurement scandal, and criminal prosecutions was a set of recommendations related to ethical conduct in acquisition and program management.

Poor Outcomes

Pressure for defense reform also grew as poor acquisition outcomes and the vulnerability of the defense acquisition system to fraud, waste, and abuse, publicized in the 1980s, raised concern about defense management. In July 1985, the time when Congress, in response to mounting dissatisfaction with defense management and organization, was considering legislation that would eventually result in the Goldwater-Nichols Act, President Ronald Reagan charged the Blue Ribbon Commission on National Defense to study and report on these and related issues. The commission, led by David Packard (of Hewlett-Packard fame and a former Deputy Secretary of Defense), produced the Packard Commission Report. In the report, the commission addressed significant defense management and execution problems, including acquisition inefficiency, cost growth, schedule delays, performance shortfalls, a lack of stability, and an unclear chain of authority. The commission considered input from both the Office of the Secretary of Defense (OSD) and the military departments in arriving at its conclusions.[14]

Budget Shortfalls

The economic recession inherited by President Reagan and the tax reductions and dramatic defense spending increases he initiated during the first half of the 1980s led to the largest budget deficits in peacetime history. Therefore, beginning in the mid-1980s, there was enormous pressure to reduce federal government spending. This pressure continued into the early 1990s and peaked after the fall of the Berlin Wall. The pressure and subsequent political negotiations resulted in a three-pronged approach to reducing spending: reductions in spending on domestic programs, reductions in defense spending beginning in the mid-1980s, and the establishment of a new tax structure that could be supported by both conservatives and liberals in the executive and legislative branches of government. In 1985, President Reagan reluctantly signed the Balanced Budget and Emergency Deficit Control Act (P.L. 99-177), popularly known as Gramm-Rudman-Hollings (GRH), which embodied the substance of the deficit reduction agreement. The measure implemented $23 billion in budget cuts across the board, split evenly between defense and domestic discretionary spending.

---

cost ballooned to $165 million. (See Chester Paul Beach, "Memorandum for the Secretary of the Navy: A-12 Administrative Inquiry," Washington, D.C., November 28, 1990.)

[13] Paul J. McNulty, *Combating Procurement Fraud: An Initiative to Increase Prevention and Prosecution of Fraud in the Federal Procurement Process*, Alexandria, Va.: U.S. Department of Justice, February 18, 2005.

[14] Packard, 1986.

Under the GRH ceiling structure, defense suffered disproportionately because much of the federal domestic program budget, particularly that covering Social Security and Medicare, was procedurally exempt from the act's provisions. The result was pressure across defense and particularly deep cuts in big-dollar weapon acquisition programs, both conventional programs (such as the C-17 cargo plane) and strategic programs (such as the Trident II missile). These programs were particularly vulnerable because moving them into planned full-scale production levels required billions in additional appropriations. As could be expected, the spreading awareness of the impending defense program budget cuts and the continued pressure by Congress to reduce spending motivated various DoD-wide efforts to curtail costs. One result was a reduction in the federal workforce, including the number of civil servants involved in the defense acquisition process.[15]

A second round of similar negotiations between President George H. W. Bush and Congress took place in the late 1980s. President Bush had inherited a $3 trillion debt, a budget proposal reflecting a $100 billion deficit, and a slowing economy. Although he proposed increases in select domestic programs, he continued the decreases in overall spending, effectively resulting in additional cuts in defense spending.

With respect to defense acquisition funding, three fundamental adjustments to defense spending in the mid- and late 1980s drastically affected the development and approval of the department's program priorities, and their influences are still felt today. The adjustments were

- the general pressure to make arbitrary reductions in labor and material costs in defense procurement, which established a new costing baseline
- the base closure and realignment process initiated by Secretary of Defense Frank Carlucci in 1987, which established a new facility size and location baseline
- the Defense Management Review (DMR) process, which promised billions of dollars in savings.

The cost and infrastructure baselines changed the "how" and "where" of material management, with obvious implications for the acquisition process. But, for the purposes of our discussion, which focuses on how deficit-reduction forces contributed to the perfect storm, we focus on the DMR.[16]

The DMR sought efficiencies throughout DoD by consolidating activities (such as by creating the Defense Finance Accounting Service, which took backroom support organizations from each defense component and consolidated them under the OSD Comptroller) and reducing support services, supply activities, and DoD headquarters (including the research, development, and procurement headquarters).

The Context in Summary

The operational problems of the U.S. military impelled Congress to change how the services selected personnel for assignment to joint duty and to revise the entire military command

---

[15] The lead authors of this paper were principal actors in determining how to allocate the programmatic and budgetary distribution of DoD reductions.

[16] Anno and Einspahr, 1988.

structure. Poor acquisition outcomes and instances of fraud hardened congressional resolve to take such steps. No single event led to the creation and passage of the Goldwater-Nichols Act, but the combination of events, and especially the ones that occurred in close succession in the latter half of the 1980s, contributed to the construction and passage of various pieces of legislation; to the internal approaches used to effect regulation and implement legislation; and, subsequently, to the continuing resolve to ensure implementation of these various legislative provisions and regulations, even in the face of emerging, unforeseen consequences.

CHAPTER THREE

The Goldwater-Nichols Act and the National Defense Authorization Act of 1987

This chapter briefly describes the passage of the Goldwater-Nichols Act, the main players in its enactment, and its key provisions. It also discusses the NDAA.

Key Players

In 1985, Senators Samuel Nunn and Barry Goldwater brought many of the issues described in Chapter Two to the attention of the Congress in a series of energetic floor speeches designed to garner political support for reform. For example, an interesting and important perspective on staff roles was articulated in views expressed by Senator Goldwater, who, in Senate floor speeches, also addressed what he perceived as the misguided financial focus of the military:

> A second consequence of this preoccupation with trying to find resources is that the military services are becoming more oriented toward business management than toward planning for and fighting a war. Our professional officer corps frequently behaves more like business managers than warriors.[1]

Senator Nunn also expounded on the issue of civilian control in the military establishment:

> [A] major problem created by the functional structure of OSD is that it encourages micromanagement of Service programs . . . [and OSD] has the tendency to get over-involved in details that could be better managed by the Services.[2]

In addition, two of Senator Nunn's major points harkened back to earlier reports on military operations:

> First, there was the lack of true unity of command, and second, there was inadequate cooperation among U.S. military services when called upon to perform joint operations. . . . The preferred advice [from the Joint Staff] is generally irrelevant, normally unread and almost always disregarded.[3]

---

[1] Barry Goldwater, "Dominance of the Budget Process: The Constant Quest for Dollars," Congressional Record, Vol. 131, No. 131, October 7, 1985, p. S12776.

[2] United States House of Representatives, Continuation of House Proceedings of October 3, 1985, No. 127; United States House of Representatives, Continuation of House Proceedings of October 4, 1985, No. 128.

[3] Nunn, quoted in Anno and Einspahr, 1988.

Senators Nunn and Goldwater together wrote on the issue of structural alignment:

> The Office of the Secretary of Defense is focused exclusively on functional areas, such as manpower, research and development, and installations and logistics. This functional structure serves to inhibit integration of Service capabilities along mission lines, and thereby, hinders achieving DoD's principal organization goal of mission integration.[4]

Representative William F. Nichols from the House of Representatives joined Senators Nunn and Goldwater in their efforts.

Key Provisions of Goldwater-Nichols

Senators Nunn and Goldwater led the effort to draft the Goldwater-Nichols Act, which was signed into law in 1986. The act made major changes in four broad areas: the chain of command and provision of military advice to the civilian leadership, the interaction of the military services, the personnel management of officers, and the acquisition of military equipment. The bill passed with wide bipartisan support: The House of Representatives vote was 383–27; the Senate's was 95–0. The act was signed into law by President Reagan on October 1, 1986.[5]

Each of the several key aspects of Goldwater-Nichols addressed in the following sections had important ramifications for DoD writ large, but their implementation in DoN had consequences whose effects were not fully understood at the time and, as is more fully discussed in Chapter Five, were likely not intended. The first two aspects served to disorient, and the latter two served to disenfranchise.

The Chain of Command and the Provision of Military Advice to the Civilian Leadership

In a key provision of the act, the process of delivering military advice to civilian authority was streamlined, and the function was centralized in the person of the Chairman of the Joint Chiefs of Staff, who became the principal military advisor to the President, the National Security Council, and the Secretary of Defense. Previously, the chiefs of the individual services had performed many associated roles; the CNO, for example, had been the advisor to the President for naval matters. The act also established the position of Vice Chairman of the Joint Chiefs of Staff, increased the ability of the Chairman to direct overall strategy, and provided greater command authority to "unified" and "specified" field commanders.[6]

Interaction Among the Military Services

The act affected service interactions by (1) diminishing the role of the service chiefs and (2) restricting the military services' operational control over forces, emphasizing instead their responsibility to support the military department secretaries in their Title 10 role to organize, train, and equip military forces for use by the Commanders-in-Chief (CINCs). The services

---

[4] Barry Goldwater and Samuel Nunn, "Defense Organization: The Need for Change," *Armed Forces Journal International*, October 1, 1985, pp. 3–22.

[5] P.L. 99-433.

[6] Unified commanders had geographical responsibilities (e.g., the Pacific area). Specified commanders had functional responsibilities (e.g., Strategic Air Command).

thus became "force providers" to the unified commanders, and their mission was to provide to the CINCs the suitably trained and equipped forces that the CINCs requested through the Joint Staff. Regardless of his service, the CINC had authority to request assets from any service through the joint system.[7]

These two changes unraveled relationships that, at least within DoN, had developed and evolved for more than 50 years. That is not to say that change is impermissible, but, in this case, there was no clear sense of the nature of the new role to be played by the service chiefs; rather, they were instructed what not to do.

The Management of Officers

Another significant but more subtle change was the direction that an officer could not receive promotion to flag rank without having completed a joint duty assignment.[8] Underlying this requirement was the perception on the part of lawmakers that the services were reluctant to send their best officers to joint duty assignments, preferring to keep them in their own ranks. Indeed, a joint duty assignment was perceived by many Navy officers as a backwater and an indication that an individual's military career was not progressing well. Officers resisted going to such assignments and, if assigned to a joint billet, tried to leave them as soon as they could. Stipulating that promotion to flag rank could not occur without a joint duty assignment ensured that the services would assign their best officers to such billets, willing or not, as a matter of necessity.

The Acquisition of Military Equipment

The Goldwater-Nichols Act specifically addressed acquisition issues, giving sole responsibility for acquisition (as part of the assignment of several "functional" areas of responsibility) to the Secretary of each military department. For example, as it pertained to DoN, Section 5014 of the act stated:

> (C) (1) The Office of the Secretary of the Navy shall have sole responsibility within the Office of the Secretary of the Navy, the Office of the Chief of Naval Operations, and the Headquarters, Marine Corps, for the following functions:
>
> - Acquisition
> - Auditing
> - Comptroller (including financial management)
> - Information management
> - Legislative affairs
> - Public affairs.[9]

---

[7] CINC (or, as they are now called, Combatant Commander) requests go to the Joint Staff, which then coordinates the delivery of requested assets with the relevant service. Requests are not automatically approved, however. For example, when Commander-in-Chief, U.S. European Command, requested Apache helicopters during the military operations in Kosovo designed to topple Serbian President Slobodan Milošević, the four services did not concur with the request. After passing out of the joint arena, the request was ultimately approved by the Secretary of Defense. See Bruce Nardulli, Walter L. Perry, Bruce R. Pirnie, John Gordon IV, and John G. McGinn, *Disjointed War: Military Operations in Kosovo, 1999*, Santa Monica, Calif.: RAND Corporation, MG-1406-A, 2002.

[8] *Flag rank* refers to generals in the Army, the Air Force, and the Marine Corps and to admirals in the Navy. Those achieving these ranks are authorized a flag whose number of stars denotes the specific rank (e.g., a brigadier general's flag has one star).

[9] P.L. 99-433.

Unlike in the other military departments, in DoN, many of these functional responsibilities were already being performed by elements of the secretariat. The word "sole" contributed to the view that the service chief was excluded from the process entirely. The act further stipulated that the Secretary designate a single organization—a service acquisition executive (SAE)—within the Secretary's office to manage the function of acquisition.

It is noteworthy that, even after the legislative changes had been passed, Senator Nunn continued to reflect on the balance of service and civilian command and control. Relevant to our investigation of the role of the CNO is Senator Nunn's concern over barriers between the military department secretary and the service chief:

> Another area that was of concern is in the consolidation of the military and civilian staffs in the military departments. The conference agreed to consolidate several functions, such as acquisition, comptroller, inspector general, and legislative liaison, under the Secretaries of the military departments and directed that the service chiefs not set up competing bureaucracies within their staffs. In the conference, *I was concerned that we not create an impenetrable wall between the staffs of the Service Secretary and the Service Chief.*[10]

Notwithstanding these concerns, the wall was built—with unfortunate consequences.

The National Defense Authorization Act for Fiscal Year 1987

The NDAA attempted to deal with several policy concerns not addressed by the Goldwater-Nichols Act. For example, it addressed the project office workload problem as represented by the excessive number of briefings that program managers (PMs) were required to give to get program approval, decreasing them to two: one to the program executive officer (PEO) and one to either the defense acquisition executive (DAE) or the SAE (depending upon the acquisition approval threshold of the program). It also addressed the need for a streamlined reporting chain from PMs to PEOs to the SAE. These and other provisions both in this act and in legislation enacted in succeeding years—the latest being the Weapon Systems Acquisition Reform Act of 2009 (P.L. 111-23)—demonstrate that the process is proceeding in a piecemeal fashion, episodically patching together solutions to address the crisis of the day. (The consequences of this approach are discussed in later chapters.) General (ret.) Lawrence A. Skantze, in a recent article in *Armed Forces Journal*, addressed one of the consequences: the inactivation in 1992 of the Air Force Systems Command.[11] He decried the resulting loss of knowledge, skills (in cost-estimating, systems engineering, contract negotiation, etc.), adult supervision, discipline, and accountability. Despite all the good intentions of acquisition reform, performance has continued to decline, and there have been concomitant slips in schedule, cost overruns, and workforce deterioration.

The next chapter describes how military acquisition was done before and after Goldwater-Nichols. This description allows us to assess the nature and scope of the changes that the legislation directed.

---

[10] Sam Nunn, statement, Conference Report, Vol. 132, No. 121, 1985, p. 10, emphasis added.

[11] Lawrence A. Skantze, "Acquisition Lost Keystone," *Armed Forces Journal*, March 2010.

CHAPTER FOUR

Acquisition Before and After Goldwater-Nichols

This chapter lays out the acquisition processes before and after the perfect storm described in the previous chapter. For convenience, the Goldwater-Nichols Act is referred to as the breakpoint, but we acknowledge that multiple influences led to a series of what might be described as tectonic shifts in acquisition processes. Furthermore, although Goldwater-Nichols was passed into law at a specific point in time, not all of its effects (or those of the other elements of the perfect storm) were felt immediately. It was several years before some of the effects were codified in DoD or defense component regulations or implemented by the Services.

This paper focuses primarily on the Navy, but we also discuss changes that occurred in the Army and the Air Force because, in some instances, those services responded to the legislation in ways that differed from the Navy, and those differences are illuminating. (Note that Marine Corps acquisition processes fall under the same DoN regulations that govern the Navy's.) In the following sections, we briefly describe the processes at the DoD level and then within the Navy, the Army, and the Air Force. Our discussion of the three services is guided by changes in service acquisition regulations, which are summarized in a table for each service.

The Office of the Secretary of Defense

Before the implementation of acquisition reforms passed in the late 1980s, and before the resulting streamlining that occurred, each military department had an acquisition organization that, relative to later years, included more stakeholders and more steps in the acquisition process. Most of the functions that now reside with the Undersecretary of Defense for Acquisition, Technology and Logistics (USD (AT&L)) were at that time assigned to the Undersecretary of Defense for Research and Engineering. Before 1986, the Secretary of Defense had overall responsibility for DoD acquisition. The Secretary of Defense and the Deputy Secretary of Defense presided over milestone decisions that are similar to those now overseen by the DAE. The most-significant change to the DoD-level acquisition regulations after Goldwater-Nichols was that many of the Secretary of Defense's acquisition decision authorities were delegated to the Undersecretary of Defense for Acquisition (USD (A)). Most significantly, the USD (A) was designated as the DAE and thus "the principal advisor to the Secretary of Defense on all matters pertaining to the Department of Defense Acquisition System."[1] Before 1987, the Deputy Secretary of Defense and various under secretaries (Research and Engineering, Policy), assistant secretaries (Acquisition and Logistics; Force Management and Personnel; Command,

---

[1] DoDD 5000.1.

Control, Communications, and Intelligence; and Comptroller), and the Director, Operational Test and Evaluation, were responsible for different aspects of the acquisition process. Further, in response to the NDAA, DoDD 5000.1 (1986) restricted the number of "management layers" between the PM and the DAE to two: the PEO and the SAE.

The Navy

Navy History and Culture

Each service has its own history and culture, and these profoundly influence how the services operate. In the case of the Navy, one of the signal differences between it and the other services appears in the very titles of the chiefs of service. Both the Army and the Air Force are headed by an individual designated as the Chief of Staff—someone who oversees the workings of a staff and is a staff officer. The head of the Navy, however, is designated CNO, which implies an individual with operational command; indeed, this aspect of CNO's office is deeply embedded in Navy history and practice. Of these service chiefs, only CNO has ever both been heavily involved in service operational matters and ultimately served as the principal advisor to the President on such matters. The point is that, historically, CNO focused on operational matters.

Until 1966, the Navy was often informally characterized as "bilinear" because CNO focused on the Navy's operational issues while SECNAV was wholly responsible for the material component, including research and acquisition elements. The tension between the military and civilian leadership of DoN over material matters was longstanding, and, historically, CNOs pushed for a greater role in acquisition matters (some even lobbied the President).[2] Organizationally, the chiefs of the Navy's material bureaus reported to the SECNAV for all material matters. In 1966, the SECNAV established the Navy Materiel Command (NMC), which was commanded by a four-star admiral with extensive operational experience and who reported to CNO. This was a major change (akin to the later tectonic shifts alluded to earlier) because it placed CNO directly in the line of material—including acquisition—issues. What was bilinear had become unilinear in that now CNO, under the direction of the SECNAV, had a direct role in the oversight of organizations involved in acquisition matters.[3]

Edwin Hooper and Thomas Hone provide rich examinations of the ebb and flow of the tide of control, which started with considerations that date all the way back to the Civil War.[4] Current discussions about increasing the authority, or even the responsibility, of Combatant Commanders simply continue the argument of who is best prepared to provide leadership in the military establishment. Shortly after Goldwater-Nichols was passed, a former SECNAV suggested that only the SECNAV could perform the role of a DoN acquisition executive

---

[2]   In a March 1934 memorandum to the SECNAV, President Franklin Roosevelt, himself a former Assistant Secretary of the Navy, wrote,

> In my judgment he [the President] would too greatly delegate this power [control of naval administration] if he delegated to the Chief of Naval Operations the duty of issuing direct orders to the bureaus and offices . . . . By this, I mean that the Chief of Naval Operations should coordinate to [sic] all repairs and alterations to vessels, etc., by retaining constant and frequent touch with the heads of bureaus and offices. But at the same time, the orders to Bureaus and offices should come from the Secretary of the Navy. (Franklin Roosevelt, "Memorandum to Secretary of the Navy," Washington, D.C., March 2, 1934)

[3]   The CNO always had influence in this area by virtue of his control over promotions and assignments, but, with the organizational realignment, he gained directive authority.

[4]   Hone, 1989.

because only the SECNAV had control of resources, could provide guidance to the service chiefs on plans and requirements, and had the political gravitas to engage both external and internal stakeholders. In his September 2008 Senate testimony regarding Goldwater-Nichols, Clark Murdock proposed an alignment that, decades after the legislation's enactment, reflects that same perspective.[5]

Acquisition Changes in the Navy

Differences in acquisition practices over time can be seen by tracing changes in Secretary of the Navy Instructions (SECNAVINSTs) that pertain to acquisition. Table 4.1 summarizes significant aspects of these instructions, showing how the instructions (presented in chronological sequence at the top of the table) affected different parts of the DoN organization.

Before the Storm

Acquisition just before the passage of Goldwater-Nichols was governed by SECNAVINST 4200.29A (1985). The wording in that instruction made SECNAV the de facto "acquisition executive" referred to in subsequent legislation and regulation. It recognized his decision authority for acquisition matters pertaining to the Navy. The instruction designated ASN for Shipbuilding and Logistics (ASN (S&L)) as the senior procurement executive and made him responsible for the performance of systems and for managing the career acquisition workforce. He was designated as the focal point for procurement and the logistical systems necessary to support the systems the Navy procured.

The instruction directed CNO to support ASN (S&L) in carrying out his duties. During this period, each of the three major warfare branches of the Navy—air, surface, and submarine—was represented by a three-star admiral on the Office of the Chief of Naval Operations (OPNAV) staff who had direct contact with the Systems Commanders for material in his warfare area. Each also had program officers who maintained a liaison with the PMs reporting to the Systems Commanders.

CNO played a direct role in the procurement process in multiple ways. His most direct role was reviewing all programs going to the SECNAV for decision. The mechanism for this review was the CNO Executive Board (CEB), on which the Vice Chief of Naval Operations also sat. As discussed below, the Systems Commanders reported material initiatives to CNO through the CEB, giving CNO a prime opportunity to engage in material management.

Although the Systems Commanders reported directly to the four-star commander of NMC, they also had reporting responsibilities to CNO; ASN (S&L); and ASN for Research, Engineering and Systems (ASN (RE&S)) in their areas of responsibility, and they were responsible for coordinating matters through NMC. The three warfare-branch vice admirals on the CNO's staff did the planning and programming for their individual warfare area systems and coordinated with NMC and the Systems Commands (SYSCOMs). Programming reviews were carried out through a CNO-chartered board. The PMs reported to the Systems Commanders through the appropriate functional SYSCOM flag officers. Figure 4.1 graphically depicts these complex relationships.

---

[5]   Clark A. Murdock, Michèle A. Flournoy, Christopher A. Williams, and Kurt M. Campbell, *Beyond Goldwater-Nichols: Defense Reform for a New Strategic Era, Phase 1 Report*, Washington, D.C.: Center for Strategic and International Studies, 2004.

Table 4.1
Acquisition Responsibilities in the Department of the Navy

	SECNAVINST 4200.29A (May 24, 1985) Regarding Procurement Executives	SECNAVINST 5430.96 (Aug 4, 1987) Assigns ASN (S&L) Responsibilities	SECNAVINST 5430.95 (Aug 5, 1987) Assigns ASN (RE&S) Responsibilities	SECNAVINST 5400.15 (Aug 5, 1991) Assigns ASN (RD&A) Responsibilities	SECNAVINST 5400.15A (May 26, 1995) Assigns ASN (RD&A) Responsibilities	SECNAVINST 5400.15B (Dec 23, 2005) Assigns ASN (RD&A) Responsibilities	SECNAVINST 5400.15C (Sep 13, 2007) Assigns ASN (RD&A) Responsibilities
SECNAV	The SECNAV is the acquisition executive and the reporting senior for the Systems Commander.	The SECNAV is the acquisition executive; some of the acquisition authorities are delegated.	Language unchanged from prior edition	The Secretary of Defense required that the MILDEP designate a single civilian office as the SAE. ASN (RD&A) is still the decisionmaker for assigned programs.	Language unchanged from prior edition	Unchanged as it pertains to the SECNAV, who is still the decisionmaker for assigned programs	Unchanged as it pertains to the SECNAV, but adds the following: Inherent in these responsibilities is the requirement to exercise good judgment, close supervision and conduct independent assessment, and the responsibility to notify DoN leadership of situations requiring their immediate attention. SECNAV is still the acquisition decisionmaker for assigned programs.

Table 4.1—Continued

	SECNAVINST 4200.29A (May 24, 1985) Regarding Procurement Executives	SECNAVINST 5430.96 (Aug 4, 1987) Assigns ASN (S&L) Responsibilities	SECNAVINST 5430.95 (Aug 5, 1987) Assigns ASN (RE&S) Responsibilities	SECNAVINST 5400.15 (Aug 5, 1991) Assigns ASN (RD&A) Responsibilities	SECNAVINST 5400.15A (May 26, 1995) Assigns ASN (RD&A) Responsibilities	SECNAVINST 5400.15B (Dec 23, 2005) Assigns ASN (RD&A) Responsibilities	SECNAVINST 5400.15C (Sep 13, 2007) Assigns ASN (RD&A) Responsibilities
ASNs	ASN (S&L) is the senior procurement executive and is responsible for (1) system performance, (2) management of the career workforce, and (3) serving as the focal point for procurement and logistics.	ASN (S&L) (1) assists SECNAV in supplying, equipping, servicing, maintaining, constructing, and outfitting ships; (2) reports to DoN acquisition executive for acquisition matters; (3) has responsibility for acquisition production and support for Navy and Marine Corps; and (4) provides such staff support as CNO and CMC each consider necessary to perform their duties and responsibilities.	ASN (RE&S) is responsible to SECNAV or the designated NAE for all DoN acquisition except shipbuilding and conversion. For matters related to research and development, ASN (RE&S) assists the NAE in executing RDT&E. The Chief of Naval Research reports to ASN (RE&S).	"ASN RDA is the Navy Acquisition Executive," a full-time role. ASN (RD&A) is responsible for (1) the development and/or procurement of systems and (2) ensuring that operational requirements are transformed within allocated resources into executable research, development, and acquisition processes.	Language unchanged from prior edition	ASN (RD&A) shall (1) manage the acquisition management structure and process, (2) recommend milestone decisions on ACAT 1D, and (3) serve as the milestone decision authority on MS ACAT 1Cs and lower.	ASN (RD&A) shall (1) exercise close programmatic oversight and provide timely reports to SECNAV and (2) independently assess programs and take action to manage program risk.

Table 4.1—Continued

	SECNAVINST 4200.29A (May 24, 1985) Regarding Procurement Executives	SECNAVINST 5430.96 (Aug 4, 1987) Assigns ASN (S&L) Responsibilities	SECNAVINST 5430.95 (Aug 5, 1987) Assigns ASN (RE&S) Responsibilities	SECNAVINST 5400.15 (Aug 5, 1991) Assigns ASN (RD&A) Responsibilities	SECNAVINST 5400.15A (May 26, 1995) Assigns ASN (RD&A) Responsibilities	SECNAVINST 5400.15B (Dec 23, 2005) Assigns ASN (RD&A) Responsibilities	SECNAVINST 5400.15C (Sep 13, 2007) Assigns ASN (RD&A) Responsibilities
Service headquarters	CNO provides support to ASN (S&L).	CNO (1) formulates and prioritizes operational military requirements, (2) conducts T&E, (3) prioritizes RDT&E, and (4) provides advice and support to SECNAV.	CNO defines the responsibilities of the principal deputy. The Director, RDT&E, reports to CNO. (CMC would be the requirements determination official if the acquisition program was a Marine Corps system rather than a Navy system.)	CNO (1) may be assigned responsibility for R&D related to military requirements and operational test and evaluation and (2) is responsible for determining requirements and establishing relative priority.[a] The "bond" between the three-star Systems Commanders and platform sponsors changed.	Unchanged	CNO (1) serves as the principal advisor to SECNAV in the allocation of resources to meet program requirements in the programming and budget processes; (2) coordinates T&E plans; (3) identifies, validates, and prioritizes the warfighting capabilities needs; and (4) determines minimally acceptable requirements.	There was an addition to the previous instruction: In coordination with ASN (RD&A), conduct an analysis of alternatives prior to the development, acquisition, and implementation of a weapon system.

Table 4.1—Continued

	SECNAVINST 4200.29A (May 24, 1985) Regarding Procurement Executives	SECNAVINST 5430.96 (Aug 4, 1987) Assigns ASN (S&L) Responsibilities	SECNAVINST 5430.95 (Aug 5, 1987) Assigns ASN (RE&S) Responsibilities	SECNAVINST 5400.15 (Aug 5, 1991) Assigns ASN (RD&A) Responsibilities	SECNAVINST 5400.15A (May 26, 1995) Assigns ASN (RD&A) Responsibilities	SECNAVINST 5400.15B (Dec 23, 2005) Assigns ASN (RD&A) Responsibilities	SECNAVINST 5400.15C (Sep 13, 2007) Assigns ASN (RD&A) Responsibilities
Acquisition, system, and technical authority commands	Systems Commanders report to four-star NMC. Systems Commanders report to CNO and to the two ASNs in their area of responsibility through NMC. PMs report to the Systems Commanders through functional flag officers.	The four-star commander, NMC, is abolished. Systems Commanders report to the DoN acquisition executive for PEO functions for all matters under the direction of the ASN (S&L) for ships.	The four-star commander, NMC, is abolished. Systems Commanders report to the DoN acquisition executive for PEO functions for all matters under the direction of ASN (RE&S) for all others.	Systems Commanders (1) manage programs other than those assigned to PEOs, (2) provide life-cycle management, (3) provide support services to the PEOs, and (4) exercise technical authority.	SYSCOMS act for and exercise the authority of the NAE to directly supervise assigned programs; they report directly to ASN (RD&A) in matters pertaining to research, development, and acquisition. The SYSCOMS report to CNO for the execution of nondevelopment, nonacquisition logistics and operating forces support responsibilities. They exercise technical authority.	Unchanged	Unchanged
Program executive oversight	The oversight of all acquisition programs was performed by the Systems Commander, as there were no PEOs.	PEOs are created and directed to report to ASN (S&L) for ship programs. PMs report to PEOs.	PEOs are created and directed to report to ASN (S&L) for ship programs. PMs report to PEOs for all programs other than ships.	PEOs report to ASN (RD&A). PMs report to PEOs.	PEOs will act for and exercise the authority of the NAE to supervise the management of assigned programs and maintain oversight of cost, schedule, and performance. PMs report to PEOs.	Unchanged	Unchanged

a CNO concurrently eliminates three-star warfare branch platform sponsors, replacing them with two-star sponsors.

Figure 4.1
Navy Acquisition Before Goldwater-Nichols

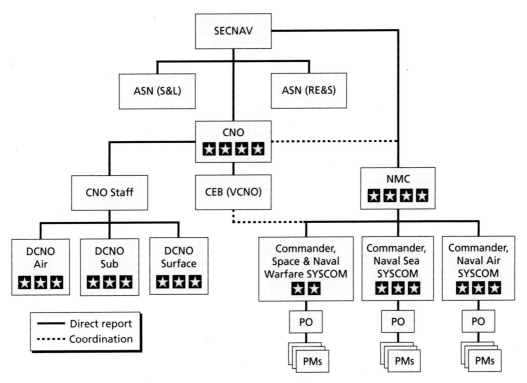

NOTE: DCNO = Deputy Chief of Naval Operations. VCNO = Vice Chief of Naval Operations.
RAND *OP308-4.1*

Although the change was not codified in Navy instructions until later, in 1985, the SECNAV abolished NMC—another of the tectonic shifts that occurred in Navy acquisition. The Chief of NMC was a four-star officer of the line who brought senior-level credibility to the material establishment and buffered the material community when needed. The disestablishment of NMC eliminated this buffer and permitted the eventual erosion of the operational credentials of the material community and of the bona fides of its proposed decisions.[6] It has been argued that NMC's ability to air differing perspectives was also the proximate cause of its disestablishment; other causes included the fact that the organization created another management layer, slowed the decision process, and ran counter to the Packard Commission's views on lines of authority.

After the Storm

DoN implemented Goldwater-Nichols in two steps. First, it designated SECNAV as the acquisition executive. Second, it attempted to use as many of the existing processes as possible to accomplish the act's intent. Both steps drew fire from the Comptroller General.[7] DoN's implementing instruction incorporated language from the Goldwater-Nichols Act regarding estab-

---

[6] For insight into the effects of this elimination of senior oversight, see Skantze, 2010.

[7] General Accounting Office, *Acquisition Reform: Military Departments' Response to the Reorganization Act: Report to Chairman, Subcommittee on Investigations, Committee on Armed Services, House of Representatives*, GAO/NSIAD-89-70, Washington, D.C.: United States General Accounting Office, June 1989.

lishment of a single organization within SECNAV's office to assume authority over the acquisition system. In doing so, the instruction stated that CNO and Commandant of the Marine Corps (CMC) "will execute their responsibilities through the resource allocation process and their input to the acquisition decision-making process."[8]

Implementing Goldwater-Nichols imposed important changes on the Navy's acquisition process. In the view of a former SECNAV, the law simply allowed too much latitude in implementation. For example, both a former General Counsel for the Navy and a former ASN (RD&A) interpreted the provision that assigned authority for the acquisition process to the military department secretaries as entirely excluding the service chiefs from the acquisition process. However, the first CNO to operate under the new provisions said that he had been unclear about his role in the acquisition process. He added that he had been advised not to get involved in acquisition decisionmaking. However, feeling that he had to be involved because he was being held "accountable" by Congress for acquisition failures, such as the A-12 aircraft program, he ignored that advice.[9]

Different interpretations also are reflected in the different forms that implementation took among the Navy, the Army, and the Air Force. Each of the military departments implemented the law differently, and all came under fire from the Comptroller General for various reasons. The common theme of these attacks was the nature of the delineation of organizational responsibilities. For acquisition, each service had PEOs reporting to the applicable military SYSCOM structure. Following the 1989 GAO report, all military departments severed the PEO structure from the SYSCOMs.[10]

With the passage of Goldwater-Nichols, the new DoN instruction, SECNAVINST 5430.96 (1987), and a companion instruction, SECNAVINST 5430.95 (1987), designated SECNAV as acquisition executive for the Navy. Thus, he not only held program decision authority but also, as acquisition executive, was responsible for the acquisition process. In support of SECNAV in that role, ASN (S&L) reported directly to SECNAV for acquisition matters. ASN (S&L) was charged with responsibility for supplying, equipping, servicing, and maintaining the Navy's equipment. He had responsibility for acquisition production and support for the Navy and the Marine Corps and for "provid[ing] such staff support as the CNO and [the Commandant] each consider necessary."[11] SECNAVINST 5430.95, dated just one day after SECNAVINST 5430.96, pertained to ASN (RE&S), who was responsible for all DoN acquisition except ship construction and conversion. He also had responsibility for matters related to research and development. In support of ASN (RE&S) in that role, the Chief of Naval Research reported to ASN (RE&S). These instructions also codified the elimination of NMC.

The most significant change occurred in the role of CNO. The new instructions divested him of acquisition responsibilities. SECNAVINST 5430.96 instead charged him with supply-

---

8 SECNAVINST 5400.15.

9 This CNO had to deal with the consequences of the unraveling of the A-12 program. In an interview for this study, he expressed the view that Congress was demanding answers from him on a range of issues with regard to the A-12 replacement program and the F-18 E/F and that, given what had occurred in the A-12 program, he had to be aware and involved in aspects of program decisionmaking, both to represent Navy interests and concerns before Congress and to be able to defend Navy resources.

10 The Army and the Air Force later gained permission from OSD to place the PEO structure back under their SYSCOMs.

11 SECNAVINST 5430.95 and SECNAVINST 5430.96.

ing, servicing, maintaining, outfitting, and logistic functions, and SECNAVINST 5430.95 directed him to formalize and prioritize requirements; conduct test and evaluation; prioritize research, development, test, and evaluation; and provide advice and support to SECNAV. Thus, CNO became responsible for determining what equipment the Navy needed but not for acquiring it. That function was now located wholly in the secretariat.

Under the provisions of SECNAVINST 5430.96, the Systems Commanders now reported to the DoN acquisition executive for all PEO matters under the direction of ASN (S&L). Similarly, the PEOs also reported to ASN (S&L).

Figure 4.2 depicts these changes. The Xs indicate the eliminations of, in this case, NMC and the dotted line between the SYSCOMs and the CEB, which still existed but had lost any approval authority. Note also that the PEOs no longer reported to the System Commanders, reporting instead directly to SECNAV, the acquisition authority.

Passage of Goldwater-Nichols did not lay to rest all acquisition issues, and all applicable organizational and process changes were not implemented immediately. For example, John Lehman, SECNAV from 1981 to 1987, designated himself as the acquisition executive but, in the view of the Secretary of Defense, that designation did not accord with the intent of the legislation, and the Secretary of Defense and his deputy pressed DoN to designate an individual ASN as the acquisition executive, eventually directing DoN to make that change. SEC-NAVINST 5400.15 (1991) codified the Secretary of Defense's direction, providing that ASN

Figure 4.2
Navy Acquisition in 1987

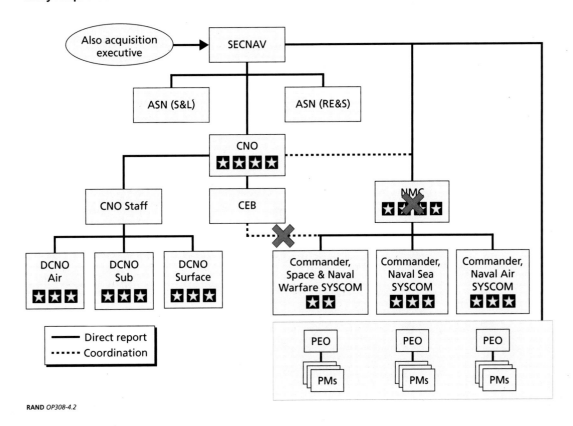

(RD&A) had a full-time role in the development and procurement of systems and in ensuring that operational requirements were transformed into executable processes. This change was another major shift in the Navy's acquisition processes.

The 1991 instruction underscored CNO's role in determining requirements and establishing their relative priority. It also indicated that CNO might be assigned responsibility for research and development matters and for operational test and evaluation, but it was clear that he could not assign himself such a role. SECNAVINST 5400.15 also codified the elimination of the "warfare branch admirals" and their relationships with the material establishment of the department.

The 1991 instruction charged the Systems Commanders with the management of programs other than those assigned to a PEO and directed them to provide support services to the PEOs. The PEOs were directed to report to ASN (RD&A), and the instruction directed PMs to report to the PEOs. The reporting line from the PEOs now ran directly to the newly named ASN (RD&A) rather than to SECNAV. SECNAV retained approval powers, but not the direct management of the process, for decisions he was empowered to make.[12] Figure 4.3 shows this continuing evolution of the Navy acquisition procedures. Key changes shown in the figure include both the changes shown in Figure 4.2 (the elimination of NMC, the reduced role of the CEB, and the formal designation of the SECNAV as the acquisition executive) and additional alterations. For example, ASN (RD&A) became the acquisition executive, and the

Figure 4.3
Subsequent Changes to Navy Acquisition Procedures

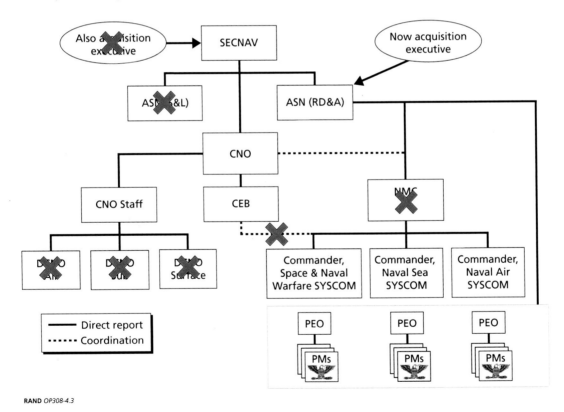

RAND OP308-4.3

---

[12] Decisions about programs that cross certain thresholds in terms of dollars for research and development and for procurement must be made at the DoD level. These are referred to as ACAT 1 decisions.

PEOs now reported to him rather than to SECNAV. ASN (AS&L) was eliminated, as were the warfare branch admirals on CNO's staff. The chain of acquisition approval now flowed directly to ASN (RD&A) rather than to SECNAV.[13]

Three instructions were published subsequent to SECNAVINST 5400.15: SECNAVINST 5400.15A (1995), SECNAVINST 5400.15B (2005), and SECNAVINST 5400.15C (2007). None changed the major responsibilities of SECNAV, CNO, or the acquisition executive, although they did elaborate on some of the functions of these positions. For example, 5400.15B designated CNO as the principal advisor to SECNAV in the allocation of resources to meet programming and budget processes. In essence, the instruction conferred on CNO the responsibility to advise SECNAV on what programmatic priorities to assign to the requirements, the development of which was his primary responsibility. He still stood outside the procurement process. SECNAVINST 5400.15C charged CNO, in conjunction with ASN (RD&A), to analyze alternatives before the development phase of a weapon system.

Instructions after 1991 also elaborated on the responsibility of the Systems Commanders and the PEOs. For example, SECNAVINST 5400.15A stipulated that the Systems Commanders would exercise the authority of the acquisition executive to supervise acquisition programs directly and, notably, would report to CNO for execution of programs that *were not* development or acquisition projects. Thus, the wall between CNO and the procurement process remained intact. PEOs were authorized to act for and exercise the authority of the acquisition executive with respect to their assigned programs and to maintain oversight of the cost and schedule performance.

Summary of Key Changes in Navy Acquisition

The process of acquiring Navy equipment changed dramatically between 1966 and 1991. Some changes were more gradual than others. The creation of the unilinear Navy took decades and crystallized with the establishment of NMC in 1966. NMC's dissolution in 1985 marked an equally significant shift. However, most key changes occurred as part of the perfect storm of events that centered on the Goldwater-Nichols legislation. Although the effects of that legislation were felt beyond the procurement process, the most critical shifts were in the roles defined for SECNAV, the ASNs, and CNO. The effect on CNO was, arguably, the greatest, since the result was his defined exclusion from the procurement process. SECNAV retained approval power but was forced to delegate responsibility for the process to one of the ASNs and to subordinate elements of the SYSCOMs. ASN (RD&A) assumed responsibilities previously carried out by SECNAV, even though, as one former SECNAV opined during an interview, only SECNAV had the responsibility and gravitas in all elements of the decision process (requirements, resources, and politics) to be able to perform the job well. The creation of the two-star PEOs eliminated the technical senior oversight that used to exist when the three-star System Commanders had oversight authority over acquisition programs and over the functions later assigned to the PEOs.

---

[13] For a relatively few years, the Undersecretary of the Navy was designated as the acquisition executive, but the duties were eventually assigned to ASN (RD&A), where they remain today.

The Army

Before the Storm

The evolution of responsibilities for the acquisition of major systems and components within the Army is summarized in Table 4.2. Like Table 4.1, this table both quotes and summarizes significant aspects of the acquisition chain of command and of organizational relationships as presented in Army Regulations (ARs)—namely, AR 70-1—from 1984 to the present.

The Secretary of the Army is responsible for all activities occurring within the department, including acquisition. Indeed, Army acquisition policy (i.e., AR 70-1) was either signed directly by the Secretary of the Army or by his order. Before implementation of Goldwater-Nichols and the NDAA provisions, the Secretary of the Army was supported by an Assistant Secretary of the Army (ASA) who was almost always designated as the Army Acquisition Executive (AAE). In 1984, the role of the AAE had been assigned to the ASA for Research, Development and Acquisition (ASA (RDA)). At the time, this individual served as an advisor to the Secretary of the Army, chaired the Army Systems Acquisition Review Council (ASARC), and decided whether acquisition programs were ready to progress past key milestones. It appears, however, that, unlike today, ASA (RDA) did not directly supervise acquisition programs or personnel. That duty resided with a uniformed officer on the Army staff, the Deputy Chief of Staff for Research, Development and Acquisition (DCSRDA). Program reviews, officer assignments, and program management assignments emanated from DCSRDA, and this three-star general, who had the staff necessary to manage the acquisition process, and worked with ASA (RDA), who had a very small staff.

The executing authority for acquisition programs resided with the Development and Readiness Command (DARCOM). The Army materiel commands, which are similar to the Navy SYSCOMs, worked for DARCOM, whose successor was the Army Materiel Command (AMC). DARCOM's commanding general reported to the Chief of Staff of the Army and the Secretary of the Army (see Figure 4.4). Thus, even though the Secretary of the Army was the acquisition decisionmaker, and although he had an assistant secretary who oversaw the acquisition system, by practice, the Chief of Staff of the Army, through DCSRDA and DARCOM, had the greatest influence over acquisition decisions. ASARC was the body that performed the highest-level review function before a Secretary of Army decision (or recommendation, as occurred in cases when the decision on a given program was being made by the Secretary of Defense). ASA (RDA) and the Vice Chief of Staff of the Army co-chaired ASARC.

After the Storm

Following the Goldwater-Nichols–era reforms, the Army reissued its acquisition regulations four times (in 1988, 1993, 1997, and 2003). In 1988, the DCSRDA position was eliminated, and ASA (RDA) was designated as the Deputy AAE and tasked to provide "principal secretariat support to the Acquisition Executive (the Secretary of the Army)."[14] The regulations issued in 1993 implemented the first structural changes that are most representative of the changes that have endured to the present day. Because the Goldwater-Nichols Act required streamlined acquisition chains of command and limited "outside" influence over acquisition activities, the acquisition chains of command were shortened to three levels for service-managed acquisitions. As mentioned earlier, the Secretary of the Army exercised overall responsibility for activities

---

[14] AR 70-1 (1988).

Table 4.2
Acquisition Responsibilities in the Department of the Army

	AR 70-1 (Aug 15, 1984) System Acquisition Policy and Procedures	AR 70-1 (Nov 12, 1986) System Acquisition Policy and Procedures	AR 70-1 (Oct 10, 1988) System Acquisition Policy and Procedures	AR 70-1 (March 31, 1993) Army Acquisition Policy	AR 70-1 (Dec 15, 1997) Army Acquisition Policy	AR 70-1 (Dec 31, 2003) Army Acquisition Policy
Service Secretary of the Army	Acquisition decisionmakers	Unchanged	Acquisition executive; acquisition decisionmaker/ recommends for assigned programs	The Secretary of the Army delegated his acquisition executive duties to ASA (RDA).	Unchanged	Unchanged
ASAs	ASA (RDA) is the AAE; the principal advisor and staff assistant to the Secretary of the Army; and responsible for overall management of research, development, and acquisition programs. ASA (RDA) serves as the Army Systems Acquisition Review Council (ASARC) decision authority, was designated as the principle board, and would make recommendations to the Secretary of the Army.	Unchanged	ASA (RDA) is the Deputy AAE and provides principal secretariat support to the AAE, to include developing policies and standards for acquisition; procurement/ contracting; technology base; program evaluation, and research, development, and acquisition planning and programming.	ASA (RDA) serves as AAE.[a]	ASA(RDA) serves as the senior procurement executive, the senior science advisor to the Secretary of the Army, and the senior research and development official for the Department of the Army.[b]	ASA (ALT) serves as the AAE, the senior procurement executive, the science advisor to the Secretary of the Army, and the senior research and development official for the Department of the Army.[c]

Table 4.2—Continued

	AR 70-1 (Aug 15, 1984) System Acquisition Policy and Procedures	AR 70-1 (Nov 12, 1986) System Acquisition Policy and Procedures	AR 70-1 (Oct 10, 1988) System Acquisition Policy and Procedures	AR 70-1 (March 31, 1993) Army Acquisition Policy	AR 70-1 (Dec 15, 1997) Army Acquisition Policy	AR 70-1 (Dec 31, 2003) Army Acquisition Policy
Military Headquarters	DCSRDA has Army General Staff responsibility for Department of the Army research, development, and acquisition activities.[d] DCSOPS has Army General Staff responsibility for the following: developing strategic concepts, plans, and broad force requirements; issuing appropriate guidance for research and development and CBTDEV programs (this guidance will include establishing and validating capability goals, material objectives, requirements development, affordability determinations, procurement of equipment, and user testing); and advising and contributing to the ASARC.	Unchanged	DCSRDA position eliminated. It is not mentioned in subsequent versions of acquisition regulations. DCSOPS has Army General Staff responsibility to develop Army policy and guidance for material requirements and combat development programs, to include validating and approving material requirements and validating capability goals and material objectives.[e]	DCSOPS will develop Army policy and guidance for material requirements and combat development programs.[f]	The Vice Chief of Staff serves as a co-chairman of the ASARC. DCSOPS will develop Army policy and guidance for material requirements and combat development programs, to include the requirements generation process and HRI processes.[g]	The Vice Chief of Staff convenes and chairs the Army Requirements Oversight Council. DCSOPS will develop Army policy and guidance for material requirements and combat development programs, to include the operational requirements generation process and HRI processes.[h]

Table 4.2—Continued

	AR 70-1 (Aug 15, 1984) System Acquisition Policy and Procedures	AR 70-1 (Nov 12, 1986) System Acquisition Policy and Procedures	AR 70-1 (Oct 10, 1988) System Acquisition Policy and Procedures	AR 70-1 (March 31, 1993) Army Acquisition Policy	AR 70-1 (Dec 15, 1997) Army Acquisition Policy	AR 70-1 (Dec 31, 2003) Army Acquisition Policy
Acquisition, system, and technical authority commands	Commanding General, U.S. Army Materiel Development and Readiness Command, is responsible in assigned areas for research, development, test, and evaluation and for acquisition and logistic support of material systems required by the Army. Specifically, Commanding General, U.S. Army Materiel Development and Readiness Command, will develop, engineer, and acquire material.	Commanding General, U.S. Army Materiel Command, is responsible as assigned for research, development, test, and evaluation and for the acquisition and logistics support of material systems required by the Army.[i]	Commanding General, U.S. Army Materiel Command, is responsible for research, development, test, and evaluation and for the acquisition and logistics support of assigned material in response to approved requirements. Commanding General, U.S. Army Materiel Command, will exercise program management oversight and decision authority over assigned programs until PEO cognizance has been established over all assigned non-PEO programs.	Commanding General, U.S. Army Materiel Command, is responsible for research, development, test, and evaluation and for the acquisition and logistics support of assigned material in response to approved requirements. Commanding General, U.S. Army Materiel Command, will plan, coordinate, and provide functional support to PEOs and PMs and will charter, supervise, evaluate, and exercise program direction and control over PMs of assigned programs.	Commanding General, U.S. Army Materiel Command, will serve as material developer for assigned programs; be responsible for research, development, test, and evaluation and for the acquisition and logistics support of assigned material in response to approved requirements; and will supervise, and evaluate assigned PMs and provide matrix support as requested by PEOs/PMs.	Commanding General, U.S. Army Materiel Command, will provide logistics and functional area matrix support as requested by PEOs/PMs.

Table 4.2—Continued

	AR 70-1 (Aug 15, 1984) System Acquisition Policy and Procedures	AR 70-1 (Nov 12, 1986) System Acquisition Policy and Procedures	AR 70-1 (Oct 10, 1988) System Acquisition Policy and Procedures	AR 70-1 (March 31, 1993) Army Acquisition Policy	AR 70-1 (Dec 15, 1997) Army Acquisition Policy	AR 70-1 (Dec 31, 2003) Army Acquisition Policy
Acquisition, system, and technical authority commands—continued	Commanding General, U.S. Army Training and Doctrine Command, directs Army combat development activities; submits requirements to HQDA for technology exploration, material development, and training devices; specifically, develops future Army concepts for doctrine, organizations, training, and material; and prepares requirements documents for new Army material and training devices.	Unchanged	Commanding General, U.S. Army Training and Doctrine Command, as the principal Army CBTDEV, formulates doctrine, concepts, organization, material requirements, and objectives; prioritizes material needs; and represents the user in the material acquisition process. Commanding General, U.S. Army Training and Doctrine Command, will coordinate and integrate the total combat developments efforts of the Army; submit requirements to HQDA for technology exploration and material and training devices based on approved operational concepts; and, specifically, will develop future Army concepts for doctrine, organizations, training, and material.	Commanding General, U.S. Army Training and Doctrine Command, as the principal Army CBTDEV, formulates doctrine and concepts; identifies requirements for future doctrine, training, leader development, organizations, and material; recommends priorities for material force modernization changes; represents the soldier in the material acquisition process; coordinates and integrates the total combat developments efforts of the Army, to include assigned strategic systems (including space-based systems) and theater or tactical interfaces to strategic systems for C3I, logistics, and management systems that support warfighting forces; and submits requirements to HQDA for technology exploration and material and training devices based on approved operational concepts.	Commanding General, U.S. Army Training and Doctrine Command, as the principal Army CBTDEV, and TNGDEV, formulates concepts; identifies requirements for future doctrine training, leader development, organizations, material, soldier, and CIE; recommends priorities for force modernization changes; represents the soldier in the acquisition process; integrates the total combat/training developments efforts of the Army; and approves Army warfighting and training requirements prior to their submission to the Secretary of the Army for prioritization and resourcing.	Commanding General, U.S. Army Training and Doctrine Command, will serve as the principal Army CBTDEV, and TNGDEV; formulate concepts; identify requirements for future DOTLMPF and CIE; recommend priorities for force modernization changes; represent the soldier in the acquisition process; and integrate the total combat and training development efforts of the Army.

Table 4.2—Continued

	AR 70-1 (Aug 15, 1984) System Acquisition Policy and Procedures	AR 70-1 (Nov 12, 1986) System Acquisition Policy and Procedures	AR 70-1 (Oct 10, 1988) System Acquisition Policy and Procedures	AR 70-1 (March 31, 1993) Army Acquisition Policy	AR 70-1 (Dec 15, 1997) Army Acquisition Policy	AR 70-1 (Dec 31, 2003) Army Acquisition Policy
Program executive oversight	Not applicable	Not applicable	PEOs will administer a defined number of assigned major and/or nonmajor programs, as approved by the AAE, ensuring that all Army agencies are responsive to the needs of the PM in achieving programmatic goals.j	PEOs administer assigned programs to ensure that all necessary support is available to achieve programmatic goals and are responsible to the AAE for programmatics (that is, material acquisition cost, schedule, and total system performance) and the planning programming, budgeting, and the execution necessary to guide assigned programs through each milestone within approved baselines.	PEOs serve as MATDEVs; are responsible for programmatics and the planning, programming, budgeting, and execution (e.g., below-threshold reprogramming authority) necessary to guide assigned programs through each milestone within approved baselines; and supervise and evaluate assigned PMs.	PEOs will serve as MATDEVs, have no other command or staff responsibilities, and only report to and receive guidance and direction from the AAE. They will be responsible for programmatics and the planning, programming, budgeting, and execution necessary to guide assigned programs through each milestone within approved baselines and established exit criteria, and they will supervise and evaluate assigned PMs. Program and project managers will supervise and evaluate assigned project and product managers.

ᵃ The AAE is responsible for administering acquisition programs according to DoD policies and guidelines. The AAE will exercise the powers and discharge the responsibilities as set forth in DoDD 5000.1 for component acquisition executives; develop and promulgate acquisition, procurement, and contracting policies and procedures; supervise and evaluate PEOs and direct-reporting PMs; and co-chair all Army System Acquisition Review Council meetings with the Vice Chief of Staff.

ᵇ Administer acquisition programs in accordance with DoD policies and guidelines; review and approve ACAT ID and ACAT 1AM programs; co-chair all ASARC meetings with the Vice Chief of Staff; serve as the milestone decision authority for Army ACAT IC and ACAT U programs.

ᶜ Exercise the procurement and contracting functions; chair the ASARC; for recapitalization, co-chair, with the Vice Chief of Staff, quarterly reviews of the overall recapitalization effort; approve, with the Vice Chief of Staff, all recapitalization baselines, baseline updates, and breach re-baselines; develop, defend, and direct the execution of the Army's acquisition policy as well as legislative and financial programs and the budget; administer acquisition programs in accordance with DoD policies and guidelines; appoint, supervise, and evaluate assigned PEOs and direct-reporting PMs; serve as the milestone decision authority for ACAT IC, IAC, and II programs; review and approve, for ACAT ID and IAM programs, the Army position at each milestone decision before the Defense Acquisition Board review or DoD CIO review (this includes the review and approval of acquisition program baselines). The AAE also approves and signs all PEO and PM charters and designates acquisition command billets.

Table 4.2—Continued

d Manage research, development, test, and evaluation activities; plan, program, and budget for acquisition of material obtained from procurement appropriations during the life cycle of material; coordinate all ASARC and Defense Systems Acquisition Review Council reviews.

e Establish and validate Army priorities throughout the PPBES, to include solutions to mission needs and research, development, and acquisition programs; approve cost and operational effectiveness analysis developed for MDAPs and ADAPs.

f This includes the requirements generation process and validating, integrating, and approving material requirements and critical operational issues and criteria for ACAT I, II, and OSD T&E oversight systems. DCSOPS will validate capability goals and material objectives; establish and validate Army priorities throughout the PPBES, to include research, development, and acquisition programs and solutions to mission needs; and establish, with HQDA participation, policy and guidance for cost, schedule, and performance trade-off analyses.

g Establish Army priorities throughout PPBES, to include research, development, and acquisition programs and solutions to mission needs; and conduct force modernization activities, develop modernization plans, and monitor the impact of force modernization planning and execution for the total Army, with the assistance of ASA(RDA) and CIO.

h Validate and integrate the review and evaluation of material requirements and critical operational issues and criteria for all ACAT programs; define and validate capability goals, material objectives, and overall force structure design; establish Army priorities throughout the PPBES, to include research, development, and acquisition programs and solutions to mission needs; co-establish, with the OASA(ALT), policy and guidance for the conduct of analyses of alternatives for major defense acquisition programs.

i Specifically, Commanding General, U.S. Army Materiel Command, will develop, engineer, and acquire material to support approved material acquisition requirements and the development of an industrial base capacity to support wartime/crisis production needs.

j Be responsible to the AAE for programmatics (that is, material acquisition cost, schedule, and performance) and the PPBE necessary to guide assigned programs through all milestones; supervise assigned project and product managers and provide the planning guidance, direction, control, and support necessary to field their systems within cost, schedule, and performance baselines.

Figure 4.4
Army Acquisition Before Goldwater-Nichols

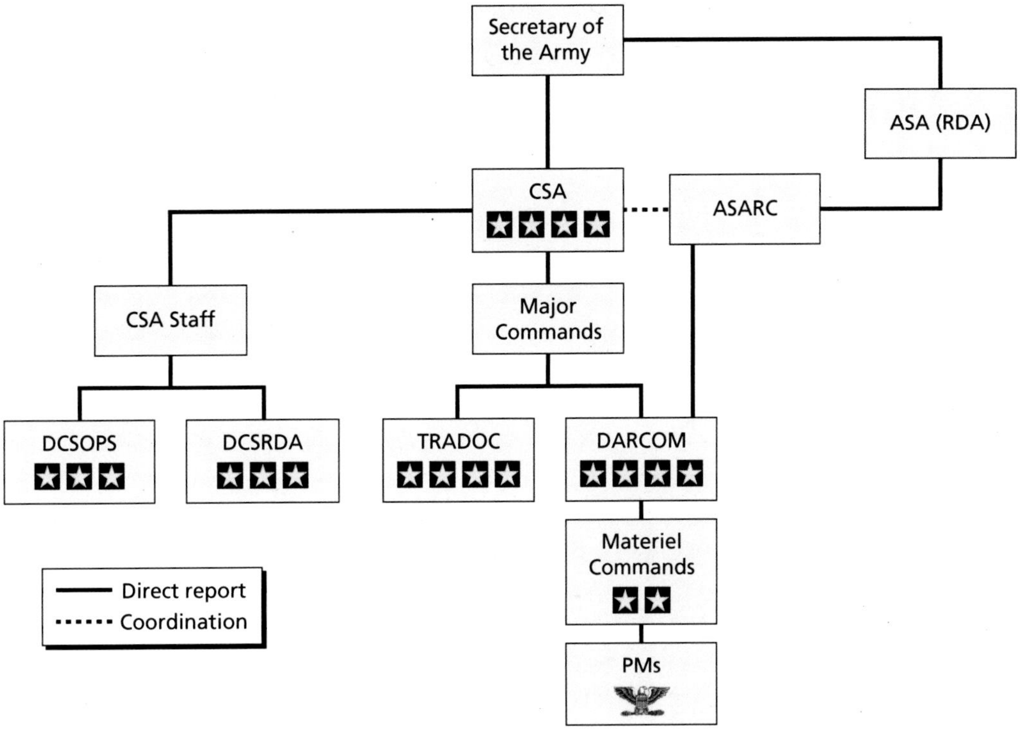

NOTE: CSA = Chief of Staff of the Army. DCSOPS = Deputy Chief of Staff for Operations and Plans.
TRADOC = Training and Doctrine Command.
RAND OP308-4.4

within the department. This revision to the Army Regulations saw the delegation of AAE responsibilities to the ASA level. The AAE role was initially assigned to ASA (RDA), whose name was later changed to ASA for Acquisition, Logistics and Technology (ASA (ALT)). Following the reforms, the AAE was more centrally positioned in the Army's acquisition process. One key aspect of that repositioning involved managing and supervising PEOs and PMs, a function that, before Goldwater-Nichols, had been performed by DCSRDA.

Also, as in the Navy, the service chief and the deputy chiefs of staff were no longer directly engaged in the acquisition process. They retained their responsibilities to produce requirements for the acquisition of new material and to develop the Program Objectives Memorandum (POM), which allocated funding to the priorities set by the President, the Secretary of Defense, and the Secretary of the Army. But, with regard to core acquisition management functions, their tasking was only to state requirements and support the PEOs and PMs. The one exception to this was the Vice Chief of Staff of the Army, who continued to co-chair the ASARC with ASA (RDA). In this role, the Vice Chief was able to represent the operational Army throughout the material acquisition process. Other commands and individuals, such as the Deputy Chief of Staff for Operations, the Training and Doctrine Command, and AMC (which had replaced DARCOM) continued to report to the Army Chief of Staff. However, the principal acquisition functions that they once managed were reorganized into a different chain of command. Figure 4.5 depicts these changes, with the Xs indicating deletions or changes.

Figure 4.5
Army Acquisition in 1993

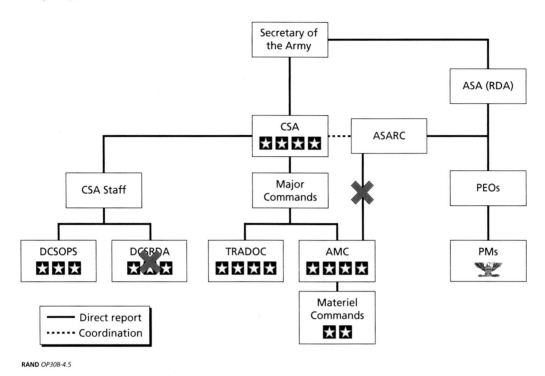

RAND *OP308-4.5*

Summary of Key Changes in Army Acquisition

The primary effects of the Goldwater-Nichols–era reforms in the Army were similar to those in the Navy. They caused the Chief of Staff of the Army and his supporting organizations, such as deputy chiefs, AMC, and subordinate Materiel Commands, to stop playing a direct role in the acquisition process. As we will see in the next section, a similar pattern emerged in the Air Force.

The Air Force

Before the Storm

The evolution of responsibilities for the acquisition of major systems and components within the Air Force is summarized in Table 4.3. Like the preceding tables for the Army and the Navy, this table quotes and summarizes significant aspects of the acquisition chain of command and organizational relationships as presented in relevant Air Force Regulations (AFRs) and Air Force Instructions (AFIs) from 1986 to the present.

The Secretary of the Air Force (SecAF) is responsible for all activities occurring within the department, including acquisition. Throughout the period under examination, SecAF was supported by an assistant secretary who was designated as the Air Force Acquisition Executive (AFAE). In 1986, this was the Assistant Secretary of the Air Force for Research, Development and Logistics. Later, the role of AFAE was assigned to the Assistant Secretary of the Air Force for Acquisition (ASAF (A)). These individuals also chaired the Air Force Systems Acquisition Review Council (AFSARC), which was the principal board that advised SecAF. (According to AFR 800-2 [1986], SecAF did not delegate his role as the milestone decision authority.) Sev-

Table 4.3
Acquisition Responsibilities in the Department of the Air Force

	AFR 800-2 (June 9, 1986) System Acquisition Policy and Procedures	AFI 63-101 (May 11, 1994) Acquisition System	AFI 63-101 (July 29, 2005) Operations of Capabilities Based Acquisition System	AFI 63-101 (April 17, 2009) Acquisition and Sustainment Life Cycle Management
Service Secretary	SecAF serves as acquisition decisionmaker.	Unchanged	Unchanged	Unchanged
Assistant secretaries	SAF/AL is designated by SecAF as the AFAE. SAF/AL is the chair of the AFSARC.	ASAF (A) is the senior corporate operating official for acquisition, the AFAE, and the SPE for overseeing Air Force acquisition activities.[a]	SAF/AQ will execute responsibilities as the senior corporate operating official for acquisition, the CAE, and the SPE for overseeing Air Force acquisition activities. He will sign all ACAT ID APBs and forward them for DAE approval; sign and approve initial APBs and any subsequent changes that constitute a re-baseline for all ACAT IC, II, and selected programs; chair Air Force Review Boards for non-space related ACAT Is, ACAT IAs, non-delegated ACAT IIs, and selected programs. SAF/US will, for space and ICBM systems, execute the CAE responsibilities assigned in the DoD series and in National Security Space Acquisition Policy 03-01.	SAF/AQ will serve as the SEA as delegated for nonspace Air Force programs; execute responsibilities as the senior corporate operating official for nonspace acquisition; and execute SEA responsibilities outlined in the DoD 5000 series for execution of nonspace Air Force acquisitions. SAF/US will serve as the SEA as delegated for Air Force space programs and as the senior corporate operating official for space system acquisition. He will execute SEA responsibilities outlined in the DoD 5000 acquisition series and in National Security Space Policy 03-01 for execution of Air Force space system acquisitions.
Military headquarters	Vice Chief of Staff will be a member of AFSARC. The DCS for Logistics & Engineering will be a member of AFSARC. The DCS for Research, Development and Acquisition will be a member of AFSARC.	No mention of acquisition responsibilities	No mention of acquisition responsibilities	The DCS for Operations, Plans and Requirements will provide oversight for Air Force planning and requirements development processes and procedures. He will collaboratively work with the acquirer, tester, sustainer, and other key stakeholders in developing operational capabilities requirements documents.

Table 4.3—Continued

	AFR 800-2 (June 9, 1986) System Acquisition Policy and Procedures	AFI 63-101 (May 11, 1994) Acquisition System	AFI 63-101 (July 29, 2005) Operations of Capabilities Based Acquisition System	AFI 63-101 (April 17, 2009) Acquisition and Sustainment Life Cycle Management
Acquisition, system, and technical authority commands	The Systems Commanders were designated line authority— i.e., management personnel who make major decisions during the acquisition of weapon systems. These personnel typically include the Secretary of Defense, SecAF, and Commander, AFSC.	AFMC supports all domestic, international, and FMS acquisition programs in which the Air Force participates; and supports the SPD by providing technical assistance, infrastructure, test capabilities, laboratory support, professional education, training and development, and all other aspects of support for AFAE, PEO, and SPD functions. Commander, AFMC, advises and assists the AFAE through formal and informal forums; may advise the AFSARC with the AFAE's written approval; and forms ad hoc assistance teams at the request of the AFAE.	AFMC will assign weapon system/program acquisition and sustainment management responsibilities to the AFMC centers through the AFMC mission assignment process; advise and assist the CAE through formal and informal forums; form ad hoc assistance teams at the request of the CAE, PEO, CD, or PM.	Commander, AFMC, will support the SAE, PEOs, and PMs by providing technical assistance, infrastructure, test capabilities, laboratory support, professional education, training and development, management tools, and all other aspects of support; support the CSAF and MAJCOM commanders by recommending phasing and adjustment of requirements to ensure operationally acceptable increments or blocks of capability are fielded in a timely manner; and support the SAE, CSAF, and MAJCOM commanders by providing direct support for requirements formulation, continuous capability and technology planning, and acquisition strategies with a focus on enhancing program success while balancing performance.
Program executive oversight	The responsibilities of the implementing command, usually AFSC, are to appoint a PM; establish a program office; and delegate program management authority and responsibility to the PM through a PM's charter.	Direct, continuous, daily interaction among the program offices, PEOs, acquisition command field activities, and headquarters staffs forms an acquisition team that ensures sound and effective acquisition practices.	PEOs will lead portfolios based on a solid business strategy designed to fulfill known capabilities needs, and they will secure necessary funding in time to meet those requirements.	PEOs will be responsible for total life cycle management of their assigned portfolios, including assigned ACAT programs, and ensure collaboration across the ILCM framework. They will be dedicated to executive management and shall not have other command responsibilities (except as waived).

a Oversees all acquisition programs through the PEO or designated acquisition commander; chairs the AFSARC; signs all ACAT IC acquisition program baselines and forwards them for DAE approval; signs and approves acquisition program baselines for all ACAT IC, II, and selected programs; supports the Air Force Chief of Staff on acquisitions. The acquisition staff supports the objectives of the AFAE and the PEOs. ASAF(A) provides all acquisition information to the PPBES.

eral members of the Air Force Chief of Staff's general staff were also assigned as members of the AFSARC. These members were the Vice Chief of Staff; the Deputy Chief of Staff (DCS) for Logistics & Engineering; and the DCS for Research, Development and Acquisition. It is interesting to note that Air Force Headquarters (AFHQ) issued program management directives that define the scope of the program being procured and provide program direction and guidance. However, the implementing command appears to have had great leeway.

The executing authority for acquisition programs resided with the "implementing command," which was designated on a program-by-program basis by the AFQH acquisition staff. One of the implementing commands named directly in AFR 800-2 was the Air Force Systems Command (AFSC). In its role as an implementing command, it was responsible for accomplishing program executive supervision in much the same way that PEOs do currently, albeit over a much larger set of programs. AFR 800-2 stated that the designated line authority for major decisions during the acquisition of weapon systems typically includes the Secretary of Defense, SecAF, and the Commander, AFSC. However, it also stated that AFHQ issues program management directives that establish programs, provide program guidance and direction, designate implementing commands, and issue the Justification of a Major System New Start to begin the acquisition process. Figure 4.6 shows Air Force acquisition before the Goldwater-Nichols Act.

Figure 4.6
Air Force Acquisition Before Goldwater-Nichols

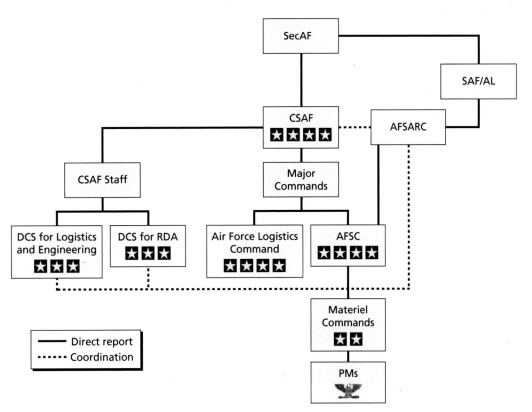

NOTE: CSAF = Chief of Staff of the Air Force. RDA = Research, Development and Acquisition.
RAND OP308-4.6

After the Storm

Following the Goldwater-Nichols–era reforms, the Air Force reissued its acquisition regulations five times (in 1990, 1993, 1994, 2005, and 2009). AFI 63-101 (1994) was the first to mention the AFAE as directly managing acquisition programs and personnel. These post–Goldwater-Nichols instructions make no mention of acquisition responsibilities within the AFHQ general staff until the most recent reissue, when AFI 63-101 (2009) tasked the Deputy Chief of Staff for Operations, Plans and Requirements with "collaboratively work[ing] with the acquirer, tester, sustainer, and other key stakeholders in developing operational capabilities requirements documents."[15]

With regard to the Air Force's materiel commands, after 1994, the instructions tasked AFMC—a successor command that, in 1994, absorbed both AFSC and the Air Force Logistics Command, eliminating one four-star position—with formally and informally advising and assisting the AFAE, PEOs, and PMs. Figure 4.7 depicts these changes.

A Comparison of the Before and After, by Department

As stated earlier, the Navy's culture differed significantly from that of her sister services and was reflected in the organization of DoN and its management structure. CNO viewed the landscape in operational terms, as befitted his title. That is, the original creation of the bilin-

Figure 4.7
Air Force Acquisition in 1994

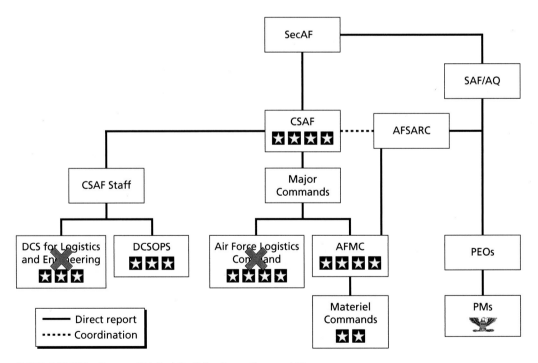

NOTE: DCSOPS = Deputy Chief of Staff for Operations and Plans.
RAND OP308-4.7

---

[15] AFI 63-101 (2009), p. 35.

ear Navy did not engage CNO immediately in the administration of the various material "bureaus" that handled the acquisition and logistical functions supporting the Navy. Ships, Ordnance, Aeronautics, and Supply and Accounts were some of the bureaus that handled such functions and reported to SECNAV. In 1966, with the creation of the Chief of Naval Material, subsequent CNOs played a greater role in the management of the material establishment and in its production. Even so, for individual CNOs who had "grown up" in the operational world—and particularly those without significant Washington, D.C., experience—dealing with material (including acquisition) matters was somewhat foreign. Furthermore, because DoN includes two services, the Navy and the Marine Corps, the DoN secretariat tended to act with greater scope than did the secretariats of the Army and the Air Force.

The term *Chief of Staff* meant, in the Army, Chief of Staff to the Secretary of the Army; in the Air Force, Chief of Staff to SecAF. As no such office existed in DoN, its secretariat was responsible for a broader set of functions that, in the other military departments, were performed in the service headquarters staffs. A couple of examples in central departmental management functions (finance and contracts) are illustrative. First, DoN's budget function reported to the ASN for Financial Management (ASN (FM)), not to either service chief. In its functions, that budget office exerted ecumenical control of the finances of the two DoN sister services and clearly worked for SECNAV. In the planning and programming processes, the Navy and the Marine Corps built their POM for SECNAV to approve, but the subsequent budget fell under SECNAV management through ASN (FM). Second, contract award approvals were managed by another ASN, depending on the item being procured. The contract award justification, called the determination and findings, had to be reviewed and approved by the appropriate office in the DoN secretariat before contracts were awarded, which meant that the secretariat now played a major role in acquisition.[16] Finally, the DoN secretariat was staffed to perform these regulatory and statutory functions and, as a result, was larger than the other military secretariats. For example, the ASN (FM) staff, including the Comptroller, at times exceeded several hundred people.

A second major difference between the Navy and her sister services was the manner in which the staffs were structured for decisionmaking. The Army and the Air Force tended to look upon issues through a functional lens. That is, when those services addressed issues, their reviews occurred at the functional level of manpower, logistics, modernization, etc. In the Navy, however, responsibility was held by three-star admirals who controlled surface, submarine, and aviation portfolios and therefore approached issues from a platform perspective. These three-star admirals also had a major voice in the requirements determination and acquisition processes before Goldwater-Nichols and had direct relationships with their three-star counterparts in the Navy SYSCOMs. When there were issues between a statement of requirements and the ability to develop acquisition programs, these issues could be resolved at the three-star level. As the Navy moved the acquisition function more fully into the DoN secretariat concurrent with the passage of Goldwater-Nichols, it abolished those three-star billets and reduced the functions they performed to the two-star level. This action, taken independently of the Goldwater-Nichols legislation, impeded the two stars' communications with the Systems Commanders, who were ultimately removed from the acquisition chain because the new PEOs

---

[16] The determinations and findings are signed, legally binding statements submitted in writing by an employee to explain or justify the method and logic used to select material, services, or suppliers when committing federal, state, or district funds for purposes of procurement of materials or services.

reported directly to ASN (RD&A). The Army and the Air Force did not experience similar concurrent changes in their structures, although the removal of the PEOs from the Systems Commanders also occurred in those departments. It is interesting to note that both the Army and the Air Force later requested and received waivers to allow the PEOs to report to what was, in essence, positions equivalent to the Navy's Systems Commanders.

Another major distinction was that the Army and the Air Force had Systems Acquisition Review Councils both before and after Goldwater-Nichols, whereas the Navy did not. In the Navy, programs wound their way through a set of SYSCOM reviews, through a two-star review in the staff of CNO, and then through the CNO-chaired CEB meeting before finally coming up for decision before SECNAV.

The ASARC and AFSARC boards were co-chaired by, respectively, the Army and Air Force service vice chiefs both before and after Goldwater-Nichols. Thus, each service chief had an important representative in councils dealing with acquisition decisions. Furthermore, our interviews with Air Force and Army retired senior general officers suggest that the principal deputy position was generally filled by a senior uniformed executive (typically a three-star general) in each of the secretariat acquisition offices and that this individual played a major role in the selection of acquisition personnel (including the management of the acquisition workforce) and had the distinct function of briefing the service chief on all matters of acquisition interest prior to his attendance in any structured meeting with the department's secretary.

With the passage and eventual implementation of Goldwater-Nichols, the Navy acquisition programs no longer went through the following organizations or reviews:

- the Systems Commanders
- the two-star CNO staff board
- the CEB.

To fill their place, the Navy Program Decision Meeting, chaired by ASN (RD&A), was created. Although CNO staff flags were invited to these meetings, the meetings were held at the behest and schedule of the ASN (RD&A), and our interviews indicate that they were poorly attended by Navy flag officers. This led to the perception that the service chief and his staff were isolated from those acquisition functions. We can thus infer that the Goldwater-Nichols Act had greater ramifications for the Navy than for the other services.

CHAPTER FIVE

How Navy Implementation Affected Acquisition

This chapter describes four major consequences of the manner in which DoN implemented the Goldwater-Nichols Act. The first is the rise of exclusive civilian control of the acquisition process and the attendant military disenfranchisement. The second is the loss of a blended workforce. The third is the separation of the "line" (i.e., naval officers who have operational assignments that lead to their promotion and success in the Navy) from the acquisition process. The fourth is both the continuing search to restore the balance and the unintended consequences of the manner in which the current DoN leadership has chosen to attempt to reintegrate the operational naval officers (line officers) into the acquisition process.

Increasing Civilian Control of the Acquisition Process: Constructing an Impenetrable Wall

During the conference leading up to enactment of Goldwater-Nichols, Senator Nunn stated that he had been "concerned that we not create an impenetrable wall between the staffs of the Service Secretary and the Service Chief."[1] In our interviews with senior Navy and OSD officials directly involved in implementing Goldwater-Nichols,[2] we found that most of these officials came to share this concern, beginning either when the act was passed or later. In fact, of the 25 former and current civilian and uniformed officials we interviewed (including those in the Air Force and the Army), all but two had no doubt that a wall had in fact been built between operational officers and acquisition officials.

In terms of our senior-level interviewees, only one—a former USD (AT&L)—believed that a minimal amount of separation between military and civilian leadership resulted from implementation of Goldwater-Nichols. Moreover, he regarded this separation as constructive in that it contributed to creative tension and led to a more efficient use of resources. In short, he believed the service chiefs could still influence acquisition decisions. Similarly, an Air Force civilian executive with a rich background in acquisition matters did not believe that the military leadership was disadvantaged by the separation of the military requirements community from their acquisition brethren. However, the remaining interviewees were much less sanguine about the nature of the outcome.

---

[1] Nunn, 1985.

[2] These interviewees included a former CNO, a former SECNAV, a former Navy General Counsel, a former ASN (RD&A), and two former Under Secretaries of Defense for Acquisition, Technology and Logistics.

According to a former principal deputy ASN (RD&A), the acquisition community eliminated roles in the acquisition process traditionally filled with staff from OPNAV. A former CNO reported that he himself felt excluded from the acquisition process, as did all the senior officers of all ranks and career fields whom we interviewed. One former USD (AT&L), who came to believe that the service chiefs and Combatant Commanders were now too far removed from the entire acquisition process, thought it essential that service chiefs become more involved in procurement planning, especially in helping to set realistic performance requirements to make trade-off decisions during program development. Both Under Secretaries of Defense (AT&L) we interviewed believed that establishing a four-star vice chief as co-chair of the Service Acquisition Board could overcome the growing divide between a military-based requirements process and a civilian-based acquisition process. In this scenario, the vice chief's role would be similar to that of the Vice Chairman of the Joint Chiefs in his role as co-chair of the Defense Acquisition Board.

The DoN leadership is not blind to this problem. It has attempted to break down the barriers between the CNO's staff and the secretariat with regard to requirements and acquisition. The Navy Gate System, initiated in SECNAVNOTE 5000 (2008), is the latest effort to link the acquisition process and requirements process. The system, shown in Figure 5.1, established a six-gate process (the gates are the numbers in the yellow circles) in which each gate represents a formal decision point at which the costs and benefits of a particular weapon system program are evaluated. The vertical dotted line that separates the first three decision gates from the last three represents a division: The first three gates are supposed to be managed on the requirements side,[3] and the last three gates are to be managed on the acquisition side.[4] That the dotted line reinforces the notion of separation and Senator Nunn's "impenetrable wall" seems to have escaped notice.

A Blended Workforce and the Engagement of Operational Officers in the Business of Acquisition

A principal motive of the Goldwater-Nichols Act was to improve the U.S. military's ability to fight in a more joint manner; consequently, joint considerations must inform not only weapon systems but also officer experience and training. All of the senior-level officials we interviewed reported that they had believed, at the time of Goldwater-Nichols implementation, that there was a need for better communication among the military departments and for more joint collaboration in operations. However, an unintended consequence of requiring officers to serve in a joint duty assignment to achieve flag or general officer rank was the migration of line officers away from the acquisition process because of the pressure of satisfying additional demands during a career whose length did not expand to accommodate these additional demands.

This migration became a particular concern in DoN because the department had, over time, maintained a blended workforce in its acquisition processes. Before Goldwater-Nichols, a mix of Navy and Marine Corps officers and technically oriented civilians were working across the material establishment. Program offices, SYSCOMs, laboratories, and field activities

---

[3] Specifically, they are managed by the Deputy CNO for Integration of Capabilities and Resources in Washington, D.C., and the Deputy Chief of Staff, Programs and Resources, CNO/CMC, in OPNAV/Headquarters, U.S. Marine Corps.

[4] Specifically, they are managed by ASN (RD&A).

Figure 5.1
The Navy Gate System

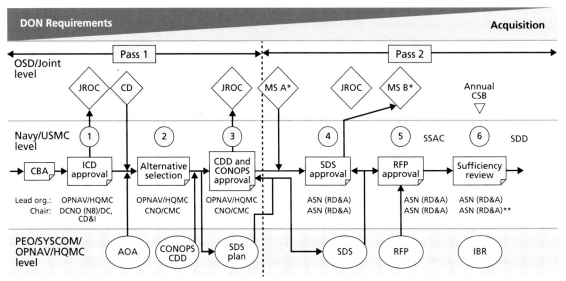

DON Requirements/Acquisition Two-Pass/Six-Gate Process with Development of a System Design Specification
(illustrated example for program initiation at Milestone A)

* DON CIO pre-certification, Investment Review Board certification, and Defense Business System (DBS) Management
   Committee approval prior to obligation of funding for a DBS program when cost > $1 million.
** Capability Production Document (CPD) review will be chaired by CNO/CMC.

AOA	Analysis of Alternative	HQMC	Headquarters Marine Corps
ASN (RD&A)	Assistant Secretary of the Navy (Research, Development and Acquisition)	IBR	Initial Capabilities Document
		ICD	Integrated Baseline Review
CBA	Capabilities-Based Assessment	JROC	Joint Requirements Oversight Council
CD	Concept Decision	PEO	Program Executive Officer
CDD	Capability Development Document	RFP	Request for Proposal
CMC	Commandant of the Marine Corps	SDD	System Development and Demonstration
CNO	Chief of Naval Operations	SDS	System Design Specification
CONOPS	Concept of Operations	SSAC	Source Selection Advisory Council
CSB	Configuration Steering Board		

(#) Gate review

SOURCE: SECNAVNOTE 5000.

RAND OP308-5.1

were generally managed by military officers who rotated into the material establishment from operational billets and brought a wealth of real-world fleet experience to these positions. They joined highly skilled engineers and scientists who, working together with former line officers, developed and procured the nation's naval weapons systems.

After the implementation of Goldwater-Nichols and the subsequent creation of the acquisition workforce there arose a formulaic career path for those whose intent was to work in acquisition. Although this path created incentives for the civilian element of the workforce, it also created significant differences between the civilian and the uniformed workforces. First, the two workforce groups had completely different chains of command and, consequently, were situated in different performance evaluation and promotion structures. The new workforce structure also demanded new educational mechanisms to prepare individuals for careers in the acquisition workforce. The Defense Acquisition University (DAU) was established with a heavy civilianized structure and outlook. The agility of acquisition was slowed by this new institutional training, and the requirement that military personnel participate in DAU courses heavily affected military career assignment and rotation.

The civilization of the workforce was accelerated by an unintended consequence of the Goldwater-Nichols emphasis on joint warfighting to satisfy promotion requirements. Before Goldwater-Nichols, officers had more time to rotate through positions related to both the operational realm and the material management process, giving those officers a deeper understanding of the civilian side of the acquisition process. With the rigid requirement of joint duty service, however, officers no longer had time to rotate between operational duty assignments and material management assignments if they wanted to achieve flag or general officer rank in an operational role. Furthermore, those who now chose to devote their energies to acquisition saw their operational experience decline relative to officers who served only in line assignments, which meant that the former lost some of their credibility when it came time to weigh in on the value of a particular performance requirement, for example. Many of the acquisition positions became "restricted" line or engineering duty officers/aerospace duty officers. As the number of officers serving in acquisition roles decreased, there emerged a sense that the acquisition process "belonged" to the largely civilian material establishment rather than to the operations or line community. Our interviews with senior Army and Air Force officers reveal the same patterns in those services. Almost to a person, our interviewees remarked on the need to create an incentive for senior line officers to serve in acquisition roles. We do not mean to imply that there is no role for restricted duty officers in the acquisition workforce. However, a blended workforce should contain officers with warfighting training and perspective to ensure that a rich mix of talent is available to the acquisition leadership.

Unintended Consequences

The Navy Gate System, with its large, structured set of meetings and briefings, needed to be established because DoN's acquisition instructions explicitly left the uniformed Navy out of the processes. Nonetheless, Senator Nunn's wall still stands—another unintended consequence of Goldwater-Nichols implementation in the Navy.

In SECNAVINST 5400 series, the applicable reference to CNO and CMC is that ASN (RD&A) "shall provide such staff support [as] each consider[s] necessary to perform his duties and responsibilities."[5] There is no mention of any other responsibility for the service chiefs. Our interviewees indicated that, when Goldwater-Nichols was passed, the uniform Navy offered a three-star deputy to ASN (RD&A). However, that offer was refused, and a senior civilian executive was installed in that position. Since then, a mix of Senior Executive Service personnel and officers with one to three stars has filled that position. But the DoN acquisition decision boards never had the uniformed Navy in any leadership position. In both the Army and the Air Force, the vice chiefs of each service at one time either chaired their acquisition decision boards or, in more-recent times, co-chaired those boards. The reason this is important is that the co-chairmanship gives the senior uniformed leadership an opportunity to demand and get information from the acquisition chain of command, starting with the PM and going up to the PEO. That information flow is important to the decision process because it provides an understanding of what is happening in the program. This insight would also allow the uniformed Navy to see the consequences of its "requirements" process and the effect of changes made in various portions of the Planning, Programming, Budgeting, and Execution (PPBE)

---

5 SECNAVINST 5400.15.

process. With this knowledge would come a shared responsibility for the end product—a most desirable effect. Another consequence of the current DoN structure is the reduced availability of operational officers who understood the acquisition process. The result is that many of the flag officers who work in the CNO's office have little or no experience with or understanding of the issues confronting acquisition programs. Therefore, the requirements process sometimes imposes unreasonable demands, and the PPBE process removes funding at critical times. In interviews, some PEOs said, "They are discussing my program in the Pentagon, and I am not even invited."

CHAPTER SIX

Conclusions, Recommendations, and Issues That Warrant Further Study

Changes that affect the culture and processes of large bureaucratic organizations are always difficult to implement. Because of their organization and their purpose, military establishments are the most difficult to change and change quickly. In the case of the defense establishment in the 1980s and 1990s, change was imposed by legislation that focused on "fixing" myriad perceived and real problems but was created without a clear understanding of the consequences of these fixes. In retrospect, it seems that many of the relevant interrelationships were not well understood. Since Goldwater-Nichols, many changes have been made in both statute and regulation to deal with "just one more" problem overlooked by previous reforms. It is interesting to speculate about what the protagonists would have done differently if they were given a glimpse of the results. But that would not be a particularly useful exercise because the problems of the day were real, and no one today is contemplating reversing what has been done. Rather, we must sift through the results of actions taken over time and see what may practically be done to address the current concerns that informed the core of our inquiry.

The various acquisition-related statutes that have been passed in the last two and a half decades reflect the changing perceptions of members of Congress about the nature of DoD's operational problems and its stewardship of the public monies and trust. Likewise, perceptions of intent have governed the promulgation of regulations to effect that legislation. Our research strongly suggests that the intent of Goldwater-Nichols was not clearly understood and that there was a significant amount of interpretation, some of it self-serving, in promulgating related military instructions, directives, and regulations. It is clear that many, despite their reservations, pressed forward anyway because the mandate for change was clear. It is also clear that DoN, because of its earlier resistance, was directed to proceed by higher authority with an even more restrictive interpretation than necessary, and did so. This letter-of-the-law approach, taken despite reservations among the leadership, resulted in a DoN implementation of the act that differed from that of the Army and the Air Force.

Senator Nunn did not seek a rigid divide between the civilian and military leadership. The Army and the Air Force have managed to avoid that divide, to a certain degree, even while facing the same statutory and directive constraints that challenge DoN. That leads us to conclude that the approach taken by DoN is more malleable than believed. Also, the de facto exclusion of offices with an operational focus from the acquisition/material management process is not healthy. Finally, to achieve the results of the process improvements discussed in the 2010 *Quadrennial Defense Review Report*, we need our best minds working together to solve problems, not sequentially engaging issues through choreographed organizational engagements. Accordingly, we present the following specific recommendations and suggest several areas that would benefit from further study.

The DoN should

- Change applicable DoN Directives to undo the isolation conveyed by the Navy Gates Process and articulate a coherent and continuing role for the service chiefs across the range of the acquisition process that is more like those of the other military departments.
- Create an acquisition oversight body co-chaired by ASN (RD&A) and the Vice Chief of Naval Operations (and, in discussions of Marine Corps systems of priority interest, the Assistant Commandant of the Marine Corps).
- Create desirable career opportunities for officers of the line in the material establishment.

Areas for further study are

- best principles and approaches to expand and rebalance the acquisition workforce to enable informed collaboration in the requirements and resources processes
- granting joint duty credit for officers in large acquisition programs, as suggested by the 2010 *Quadrennial Defense Review Report* (which recommends "[r]ecognizing joint experience whenever and wherever it occurs"[1])
- appropriate changes in officer assignments to create enhanced senior officer opportunities in acquisition.

In closing, we note that the wall between the service secretary and the service chief remains a concern among policymakers to this day. The following recommendation appeared in a recent House Armed Services Committee document:

> **The Department [of Defense] and Congress should review and clarify the Goldwater-Nichols Act's separation between acquisition and the military service chiefs to allow detailed coordination and interaction between the requirements and acquisition processes and to encourage enhanced military service chief participation in contract quality assurance.**

> The Panel is concerned that the divide established in the Goldwater-Nichols Act between acquisition and the military service chiefs has become so wide that it hinders both the acquisition and requirements process. While the fundamental construct in the Goldwater-Nichols Act, correctly assigned lead responsibility for acquisition to the Department's civilian leaders, the act should be clarified to ensure that the requirements process that must coordinate with all categories of the defense acquisition system freely interacts with the acquisition process. The service chiefs should also be given greater authority and responsibility to oversee contract quality assurance especially for contracts that are highly operational in nature.[2]

---

[1] Department of Defense, *Quadrennial Defense Review Report*, Washington, D.C.: Department of Defense, February 2010, p. 54.

[2] United States House of Representatives, House Armed Services Committee Panel on Defense Acquisition Reform Findings and Recommendations, Washington, D.C., March 23, 2010, emphasis in the original.

On July 27, 2010, the Quadrennial Defense Review Independent Panel chaired by former Secretary of Defense William J. Perry and former National Security Advisor Stephen J. Hadley stated, "The Panel believes that the fundamental reason for the continued under performance in acquisition activities is fragmentation and accountability for performance."[3] The main thrust of this paper is precisely the same: The military requirements community and the civilian-run acquisition community need to come together to share the authority and responsibility for the performance of the requirements and acquisition system in DoD. Like President Reagan, the House Committee and the Quadrennial Defense Review Independent Panel believe that the wall should be torn down.

---

[3]   Stephen J. Hadley and William J. Perry, *The QDR in Perspective: Meeting America's National Security Needs in the 21st Century—The Final Report of the Quadrennial Defense Review Independent Panel*, July 27, 2010, p. 84.

Former Positions of Interviewees

Our interviewees held the following positions:[1]

- Army Director of Requirements
- Assistant Commander, Test and Evaluation, Naval Air Systems Command
- Assistant Deputy Chief of Staff for Installations and Logistics, Marine Corps
- Assistant Secretary of the Army for Acquisition, Logistics and Technology (Army Acquisition Executive)
- Assistant Secretary of the Army for Research, Development and Acquisition
- Assistant Secretary of the Navy for Research, Development and Acquisition
- Chief of Naval Operations
- Commander, Aeronautical Systems Division at Wright-Patterson Air Force Base
- Commander in Chief, U.S. Atlantic Command
- Commander, Naval Air Systems Command
- Commander, Naval Air Warfare Center, Weapons Division (2)
- Commander, U.S. European Command
- Commanding General, Army Materiel Command
- Commanding General, Army Operational Test and Evaluation Command
- Deputy Assistant Secretary of Defense (Materiel Acquisition)
- Deputy Chief, Naval Materiel Command
- Deputy Chief of Naval Operations for Fleet Readiness and Logistics
- Deputy Chief of Staff for Plans and Operations
- Deputy Chief of Staff for Research, Development and Acquisition
- Deputy for Systems Management and Horizontal Technology Integration
- Deputy Program Executive Officer, Tactical Aircraft Programs, Naval Air Systems Command
- Director, Acquisition Excellence Aeronautical Systems Center
- Director, Army Acquisition Corps (3)
- Director, Aviation Plans and Requirements Division
- Director of Air Force Operational Requirements
- Director of Defense Research and Engineering (Electronics)
- Director of Ship Research and Development, Naval Sea Systems Command

---

[1] The number in parentheses after certain positions indicates that we interviewed that number of people who held that particular position.

- Director of Tactical Programs in the Office of the Assistant Secretary of the Air Force for Acquisition
- Executive Director, Naval Air Systems Command
- Executive in Residence, Defense Acquisition University
- Military Deputy to the Assistant Secretary of the Army for Acquisition, Logistics and Technology (2)
- Military Deputy to the Assistant Secretary of the Army for Research, Development and Acquisition
- Naval Air Systems Command Assistant Commander for Test and Evaluation, and for Shore Installation Management
- Navy Program Executive Officer for Ships (2)
- Navy Program Executive Officer for Submarines
- Principal Deputy General Counsel, Department of Defense
- Principal Deputy General Counsel, Department of the Navy
- Principal Deputy, Assistant Secretary of the Navy for Research, Development and Acquisition
- Program Executive Officer, Air Force
- Program Executive Officer, Army (2)
- Program Executive Officer, Navy (3)
- Program Executive Officer, Tactical Aircraft Programs
- Program Manager, Air Force
- Program Manager, Army
- Program Manager, Navy (5)
- Secretary of the Air Force
- Secretary of the Navy (2)
- Under Secretary of Defense for Acquisition, Technology and Logistics (2)
- Vice Chairman, Joint Chiefs of Staff/Chair, Joint Requirements Oversight Council
- Vice Chief, Naval Sea Systems Command
- Vice Chief of Naval Operations
- Vice Chief of Staff, Air Force

References

Anno, Stephen E., and William E. Einspahr, "The Grenada Invasion," in Stephen E. Anno and William E. Einspahr, *Command and Control Lessons Learned: Iranian Rescue, Falklands Conflict, Grenada Invasion, Libya Raid*, Maxwell Air Force Base, Ala: Air University Press, 1988.

Assessment Panel of the Defense Acquisition Performance Assessment Project, *Defense Acquisition Performance Assessment: Executive Summary*, December 2005.

Beach, Chester Paul, "Memorandum for the Secretary of the Navy: A-12 Administrative Inquiry," Washington, D.C., November 28, 1990.

Cheney, Dick, *Defense Management Report to the President*, Washington, D.C.: Department of Defense, 1989.

Committee on Armed Services, House of Representatives, "Background Material on Structure Reform of the Department of Defense," 99th Congress (2nd Session), Washington, D.C.: U.S. Government Printing Office, 1986.

Department of Defense, *Operation of the Defense Acquisition System*, DoDI 5000.02, Washington, D.C.: Department of Defense, multiple iterations.

———, *Major System Acquisitions*, DoDD 5000.1, Washington, D.C.: Department of Defense, March 12, 1986.

———, *Quadrennial Defense Review Report*, Washington, D.C.: Department of Defense, February 2010.

Department of the Air Force, *System Acquisition Policy and Procedures*, AFR 800-2, Washington, D.C.: Department of the Air Force, June 6, 1986.

———, *Acquisition System*, AFI 63-101, Washington, D.C.: Department of the Air Force, May 11, 1994.

———, *Operations of Capabilities Based Acquisition System*, AFI 63-101, Washington, D.C.: Department of the Air Force, July 29, 2005.

———, *Acquisition and Sustainment Life Cycle Management*, AFI 63-101, Washington, D.C.: Department of the Air Force, April 17, 2009.

Department of the Army, *System Acquisition Policy and Procedures*, AR 70-1, Change 1, Washington, D.C.: Department of the Army, August 15, 1984.

———, *System Acquisition Policy and Procedures*, AR 70-1, Washington, D.C.: Department of the Army, November 12, 1986.

———, *System Acquisition Policy and Procedures*, AR 70-1, Change 1, Washington, D.C.: Department of the Army, October 10, 1988.

———, *System Acquisition Policy and Procedures*, AR 70-1, Change 1, Washington, D.C.: Department of the Army, March 31, 1993.

———, *System Acquisition Policy and Procedures*, AR 70-1, Change 1, Washington, D.C.: Department of the Army, December 15, 1997.

———, *Army Acquisition Policy*, AR 70-1, Washington, D.C.: Department of the Army, December 31, 2003.

Department of the Navy, *Department of the Navy Procurement Executives*, SECNAVINST 4200.29A, Washington, D.C.: Department of the Navy, May 24, 1985.

————, *Assignment of Responsibilities to the Assistant Secretary of the Navy (Shipbuilding and Logistics)*, SECNAVINST 5430.96, Washington, D.C.: Department of the Navy, August 4, 1987.

————, *Assignment of Responsibilities to the Assistant Secretary of the Navy (Research, Engineering and Systems)*, SECNAVINST 5430.95, Washington, D.C.: Department of the Navy, August 5, 1987.

————, *Department of the Navy Research, Development and Acquisition Responsibilities*, SECNAVINST 5400.15, Washington, D.C.: Department of the Navy, August 5, 1991.

————, *Department of the Navy Research, Development and Acquisition, and Associated Life Cycle Management Responsibilities*, SECNAVINST 5400.15A, Washington, D.C.: Department of the Navy, May 26, 1995.

————, *Department of the Navy Research, Development and Acquisition, and Associated Life-Cycle Management Responsibilities*, SECNAVINST 5400.15B, Washington, D.C.: Department of the Navy, December 23, 2005.

————, *Department of the Navy (DoN) Research and Development, Acquisition, Associated Life-Cycle Management, and Logistics Responsibilities and Accountability*, SECNAVINST 5400.15C, Washington, D.C.: Department of the Navy, September 13, 2007.

————, *Department of the Navy (DoN) Requirements and Acquisition Process Improvements*, SECNAVNOTE 5000, Washington, D.C.: Department of the Navy, 2008.

DoD Commission, *Report of the DoD Commission on Beirut International Airport Terrorist Act*, Washington, D.C.: Department of Defense, 1983.

Doty, Joseph P., *Urgent Fury—A Look Back . . . A Look Forward*, Newport, R.I.: Naval War College, 1994.

Ferraro, Peter J., *Beirut, Lebanon: 24th MAU, May–Dec 1983*, Decatur, Ga.: Marine Corps University Command and Staff College, 1997.

General Accounting Office, *Acquisition Reform: Military Departments' Response to the Reorganization Act: Report to Chairman, Subcommittee on Investigations, Committee on Armed Services, House of Representatives*, GAO/NSIAD-89-70, Washington, D.C.: United States General Accounting Office, June 1989.

Goldwater, Barry, "Dominance of the Budget Process: The Constant Quest for Dollars," *Congressional Record*, Vol. 131, No. 131, October 7, 1985.

Goldwater, Barry, and Samuel Nunn, "Defense Organization: The Need for Change," *Armed Forces Journal International*, October 1, 1985, pp. 3–22.

Hadley, Stephen J., and William J. Perry, *The QDR in Perspective: Meeting America's National Security Needs in the 21st Century—The Final Report of the Quadrennial Defense Review Independent Panel*, July 27, 2010, p. 84.

Holloway, James L., *Special Operations Review of Iranian Hostage Rescue Mission, Joint Chiefs of Staff*, Washington, D.C., August 23, 1980.

Hone, Thomas C., *Power and Change: The Administrative History of the Office of the Chief of Naval Operations, 1946–1986*, Washington, D.C.: Naval Historical Center, 1989.

Hooper, Edwin B., *The Navy Department: Evolution and Fragmentation*, Washington, D.C.: Naval Historical Foundation, 1978.

Joint Defense Capabilities Study Team, *Joint Defense Capability Study: Improving DoD Strategic Planning, Resourcing and Execution to Satisfy Joint Capabilities*, Washington, D.C.: Department of Defense, 2004.

Locher, James R. III, "Has It Worked?: The Goldwater-Nichols Reorganization Act," *Naval War College Review*, Vol. LIV, No. 4, 2001, pp. 95–114.

McMaster, H. R., *Dereliction of Duty: Lyndon Johnson, Robert McNamara, the Joint Chiefs of Staff, and the Lies That Led to Vietnam*, New York: Harper-Collins, 1997.

McNulty, Paul J., *Combating Procurement Fraud: An Initiative to Increase Prevention and Prosecution of Fraud in the Federal Procurement Process*, Alexandria, Va.: U.S. Department of Justice, February 18, 2005.

Murdock, Clark A., Michèle A. Flournoy, Christopher A. Williams, and Kurt M. Campbell, *Beyond Goldwater-Nichols: Defense Reform for a New Strategic Era, Phase 1 Report*, Washington, D.C.: Center for Strategic and International Studies, 2004.

Nardulli, Bruce, Walter L. Perry, Bruce R. Pirnie, John Gordon IV, and John G. McGinn, *Disjointed War: Military Operations in Kosovo, 1999*, Santa Monica, Calif.: RAND Corporation, MG-1406-A, 2002. As of June 25, 2010:
http://www.rand.org/pubs/monograph_reports/MR1406/

Nunn, Sam, statement, Conference Report, Vol. 132, No. 121, 1985.

Packard, David, *President's Blue Ribbon Commission on Defense Management, A Quest for Excellence: Final Report to the President*, Washington, D.C., June 30, 1986.

Public Law 99-177, Balanced Budget and Emergency Deficit Control Act, December 12, 1985.

Public Law 111-23, Weapon Systems Acquisition Reform Act of 2009, May 22, 2009.

Public Law 99-433, Goldwater-Nichols Department of Defense Reorganization Act of 1986, October 1, 1986.

Public Law 99-961, National Defense Authorization Act for Fiscal Year 1987, November 14, 1987.

Record, Jeffrey, *The Wrong War: Why We Lost in Vietnam*, Annapolis, Md.: Naval Institute Press, 1998.

Roosevelt, Franklin, "Memorandum to Secretary of the Navy," Washington, D.C., March 2, 1934.

Skantze, Lawrence A., "Acquisition Lost Keystone," *Armed Forces Journal*, March 2010.

United States House of Representatives, Continuation of House Proceedings of October 3, 1985, No. 127.

———, Continuation of House Proceedings of October 4, 1985, No. 128.

———, House Armed Services Committee Panel on Defense Acquisition Reform Findings and Recommendations, Washington, D.C., March 23, 2010.

Wetterhahn, Ralph, *The Last Battle: The Mayaguez Incident and the End of the Vietnam War*, New York: Carroll and Graf, 2001.